The Secret Life of the HUMAN BODY

Uncover the hidden workings of your body

JOHN CLANCY

An Hachette UK Company
www.hachette.co.uk

First published in Great Britain in 2018 by Cassell,
an imprint of Octopus Publishing Group Ltd
Carmelite House
50 Victoria Embankment
London EC4Y 0DZ
www.octopusbooks.co.uk

Copyright © Octopus Publishing Group Ltd 2018

Edited and designed by Tall Tree Limited

ISBN 978 1 84403 978 4

A CIP catalogue record for this book is available
from the British Library.

Printed and bound in China

10 9 8 7 6 5 4 3 2 1

Contents

Introduction

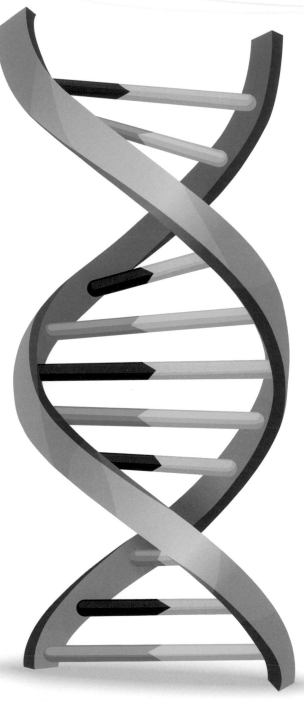

You are about to begin the fascinating, dynamic study of the human body. For thousands of years, people have been observing and investigating the human body, trying to comprehend how it works when it is healthy and what goes wrong during ill health.

Discovering the Body

Before the end of the 16th century, medics had to rely on studying the human body armed only with the naked eye, and they often relied on "gut instinct" to treat disease and illness. Thankfully, since then, we have made great strides forward in our knowledge base of the human body. Inventions that have made this possible include the light and electron microscopes in the 20th century. The invention of the electron microscope was responsible for a major breakthrough in our knowledge in 1953, when Watson and Crick made public that they had found "the secret of life" and announced the structure of deoxyribonucleic acid (DNA). The current use of medical imaging techniques, including X-rays, ultrasound, three-dimensional Computerized Tomography (CT) and Magnetic Resonance Imaging (MRI) scans, have progressed our knowledge base of the human body.

Arguably, the greatest innovation recorded to date was in 2003, with publication of the Human Genome Project. This identified the exact position of each of the 25,000 genes within our

The complex double-helix shape of a DNA molecule was identified as the genetic blueprint from which a human being is made and is responsible for identifying differences between individuals.

46 chromosomes. As such, it led to the first individual having their entire DNA sequence uploaded onto the Internet in 2007. It is predicted that in the near future our medical records will include our own genome, so treatments can be personalized to you.

What's Next, You Might Ask?

Well, currently, the research regarding the human genome is ongoing, and almost every week, a gene sequence has been associated with either a healthy or diseased trait. The natural evolution of genomic knowledge is to identify the products of gene activity, so a group of scientists became involved in the Human Proteomics Project. This involves studying the gene products, that is enzymes, their structure and their functional activity. These studies will provide more information as to what is happening in the cell. This has applications for drug design, tailoring them to the individual. There is also a group of researchers investigating how environmental factors affect gene-enzyme activity in health and disease. This could reveal how risk factors such as smoking, stress, microbes and other environmental hazards can affect the human body. All in all, we are experiencing a very exciting time in the study of the human body, which will no doubt provide a longer and healthier life span for us humans.

Armed with this advanced technology, we can now explore the microscopic world inside the human body. By studying this book you will discover an amazing collection of secrets about your body, such as how the it is made up of trillions of microscopic structural and functional units called cells. Each cell can be regarded as the "basic unit of life", since it is the smallest part of the body capable of performing all the basic needs necessary for your survival. Cells can digest food, generate energy, move, respond to stimuli, grow, excrete and reproduce. To support these basic needs, cells contain organelles or "little organs" that carry out specific activities. You will see from chapters 1 and 11 that the genes we inherit from our parents are the controllers of the secret chemistry associated with our cells' activities and hence the health of the human body. They are also involved when disease strikes.

The study of the human body involves several branches of science: human biology, chemistry, physics, mathematics, psychology and sociology. Each contributes an understanding of how the body functions in health, during times of exercise, illness, pain, distress, trauma and surgery. However, it must be stressed that human beings are biological organisms. The two interwoven branches of science covered in this book that will help you understand the human body are anatomy, that is how your body is organized structurally, and physiology, that is how it functions. Identifiable within these is a concept referred to as homeostasis.

Homeostasis: "A Happy Healthy Body"

Homeostasis refers to the automatic actions within the body, which are necessary to maintain the "healthy" consistent state of the body's environment, despite changes in the environment outside the body. Collectively, anatomical structure, physiological function (or activities) and the maintenance of homeostasis enable the cells to function and perform the basic needs of life necessary for a "happy healthy body".

If homeostasis provides a secret basis for health in the human body, then illness occurs when there is a failure of the automatic actions within the body, which are necessary to maintain the "normal" healthy status quo. This book highlights some of the more common illnesses of the human body. However, it cannot be overstated that illness may be classed according to the primary disorder, such as a stomach or lung problem, a tumour of a particular tissue, such as the breast, or as being due to an infection, such as pneumonia. Nonetheless, all will have consequences elsewhere in the body, other than those involved in the primary disturbance, and so the healthcare we receive when we are "ill" may be directed at symptoms apparently far removed from the primary problem. For example, these may include relieving constipation in a patient who has a colorectal tumour.

We Are All Different

We only have to look at other people around us to recognize that genetic variation dictates not only the colour of our hair, eyes and skin, but also how individuals respond to stress, the diseases to which we are susceptible and even how we react to different medicines. However, despite this variation, we are all built to the same basic template, with the same body systems, and our cells working the same basic way. It is with this in mind that this book takes a systems approach, since it allows greater understanding of how the body works as a whole. The human body is made up of a number of different systems, each with their own separate function. These systems are linked together through the circulatory and lymphatic systems and communicate with each other through the nervous and endocrine systems. Together, all the systems allow the body to move, explore and interact with the environment, and to carry out activities which are vital to health and our survival. A deterioration in one system, however, leads to other systems being affected, and ultimately, when the systems cannot operate anymore, this can or this will lead to our death.

With more than seven billion people on the planet, the amount of genetic variation is truly staggering. Even so, we humans all follow the same body blueprint allowing medics to simplify treatment for a whole host of diseases.

Chapter 1
CELLS, TISSUES AND BODY STRUCTURE

The Cell

Cells make up all living things (including you!), and that's why we refer to them as the "building blocks of the body".

Cells are also called the "basic units of life" because they are the smallest parts of the body that are capable of performing the basic needs of life. Cells can digest food, generate energy, move, respond to stimuli, grow, excrete and reproduce. Just as the body has organs to perform specialized functions, cells have small component parts called organelles, or "little organs", that have unique structures and perform specific roles within the cell.

Control Substances

The production of enzymes is the cell's main job. This is because enzymes produce all the components that the cells need, such as proteins, complex carbohydrates and complex lipids. Enzymes also break down things that the cell doesn't need so that you can maintain a happy, healthy body (homoeostasis). They are also responsible for cell division in growth, and the repair and regeneration of cells when they are damaged and need to be replaced.

Organizing the Body

How do you get from a cell to a working human body? It's a process of organization. Cells within the body differentiate and specialize to form tissues that carry out a particular function. These tissues work together to form organs, such as the stomach, which is an organ made up of different tissues. Organs that work together form organ systems, such as the digestive system, which work with other organ systems to make an entire organism, the human body.

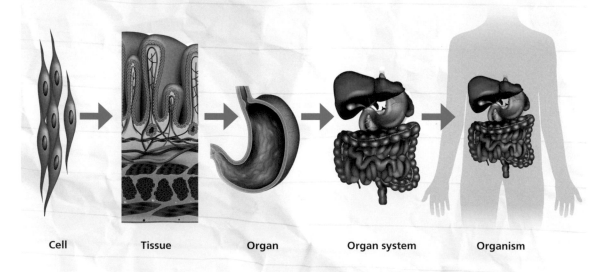

Cell   Tissue   Organ   Organ system   Organism

The Cell Factory

The different parts and activities of the cell can be compared to a factory. Many different and important tasks are completed within a factory building, just like the varied parts of a cell.

The nucleus is the main office of the factory, where all the cell's activity is controlled. It contains genes (made of DNA) that give the instructions for making proteins, called enzymes. The nucleus is surrounded by a nuclear membrane, which controls what goes in and out of it.

Cell Membrane

Similar to the fence around a factory, the cell membrane regulates the entry and exit of things into and out of the cell and provides protection against invasion. It also contains specific proteins, called antigenic markers, which give the cell its identity. For example, the antigenic markers of heart cells and kidney cells differ from one another

and they also differ in individuals, since they are determined by an individual's genetics. It is these markers that are important in matching up donors and recipients so that successful organ transplants can take place.

Cytoplasm

The main building of the factory, the cytoplasm contains everything within the cell, except components of the nucleus.

Nucleoplasm

Equivalent to the main office of the factory, the nucleoplasm houses components, such as the chromosomes of the cell's nucleus.

Typical cell

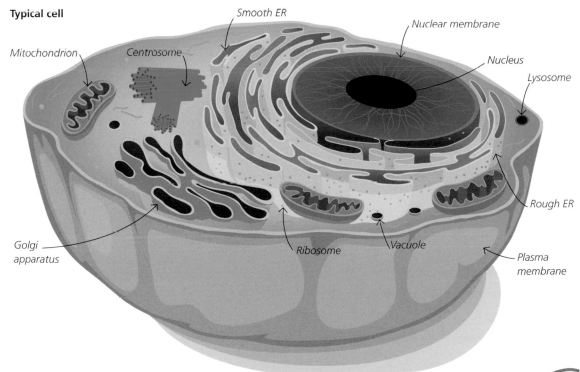

Mitochondrion · Centrosome · Smooth ER · Nuclear membrane · Nucleus · Lysosome · Rough ER · Golgi apparatus · Ribosome · Vacuole · Plasma membrane

Mitochondria

These are like the power generator of the factory, supplying the electrical energy for the factory's machinery to work. In the cell, the mitochondria break down food (mainly sugars) with oxygen in a process called aerobic respiration. This provides energy in the form of adenosine triphosphate (ATP), which drives chemical reactions. It also releases heat that maintains body temperature and provides the best conditions for enzyme activity. This is why the mitochondria are often called the "powerhouse" of the cell. The other end products of aerobic respiration – carbon dioxide and water – are important as they maintain the pH (acidity or alkalinity) of the cell, which also optimizes conditions of enzyme activity.

Endoplasmic Reticulum (ER)

The endoplasmic reticulum (ER) is a network of membranes found throughout the cell and connected to the nucleus. It is equivalent to pipelines in a factory through which supplies are moved. The rough endoplasmic reticulum has ribosomes attached to its outer surface. Ribosomes are similar to the machines on the assembly line that put the parts together to make the protein products (enzymes) of the factory. The smooth endoplasmic reticulum lacks ribosomes and is involved in the storage and synthesis of lipids (fatty acid compounds).

Golgi Bodies

Golgi bodies are similar to the packaging department, where products (carbohydrates, proteins and lipids) are packed together with enzymes, either to be used in the cell or exported.

Lysosomes

These are similar to the cleaning crew of the factory. The lysosomes break down (with enzymes): lipids; proteins; carbohydrates; foreign bodies that have entered the cell; worn out parts of the cell and even the cell itself, when it is no longer functioning at its best. That is why the lysosome is sometimes called the "suicide bag" of the cell.

The cell membrane acts like security, controlling what goes into and out of the cell.

Centrosome

The centrosome is like the financial controller of the factory, deciding whether to enlarge the factory or to reduce its size depending upon its financial position. In a similar way, centrosomes are important in mitosis – chromosomal duplication required for cell multiplication – and in meiosis – chromosomal reduction in sperm and egg formation. Centrosomes develop microtubules (similar to little muscles), which pull apart duplicating chromosomes in mitosis to ensure daughter cells have identical numbers of chromosomes to the parent cell. In meiosis, centrosomes split chromosomes in two, before the sex cells are produced. This makes sure that the daughter cells have half the chromosomes of the parent cell (see Making Sex Cells on page 17).

Vacuoles

Like a factory's storage, vacuoles in the cell store water, salts, proteins, lipids and carbohydrates.

The factory is open 24/7 (even when you're sleeping) and its work is never done!

Cells, Tissues and Body Structure

Genes: "the Secret Code of Life"

Genes are the controllers of all the cell's chemical reactions (collectively called metabolism) and indirectly affect enzyme production. Enzymes are biological catalysts and are therefore of fundamental importance in the human body. This is because enzymes speed up chemical reactions in our cells, which are crucial for a "healthy, happy body".

Genes help produce all the products of the cell that give you your observable traits, such as hair, eye and skin colour. They also control activities of the body, such as digestive enzymes, tears, blood clotting proteins, antibodies, hormones and even new cells. Genes are commonly referred to as the "code of life" and enzymes as the "key chemicals of life", because body function and even creation would not be possible without them.

Under Lock and Key

An enzyme only accepts a molecule (the substrate) with the correct shape, just like a key will only fit into a certain lock. This "lock and key" principle applies to various chemical interactions within the body, such as in the combination of an antibody (blood protein) with an antigen (foreign substance) during an immune response. A lack of a particular enzyme will prevent a certain reaction from occurring, while an excess of the enzyme could cause the reaction to proceed too quickly. Controlling the level of enzyme synthesis is therefore essential in sustaining homeostasis or the balance of chemicals within the human body.

Enzymes produced by a cell are normally for the cell's own use – which is like giving a present to itself! In addition to enzymes that are common to processes in all cells, there are also specific enzymes that relate to the role of a particular cell type. Enzymes may become detectable in blood if the cells are damaged and those that are specific to the cell type may then be isolated and used to diagnose certain conditions. For example, the appearance in blood of the enzyme lactic dehydrogenase (LDH) is an indicator of heart or liver cell damage, which then allows doctors to diagnose conditions like angina, myocardial infarction (heart attack) and liver disease.

The lock and key diagram

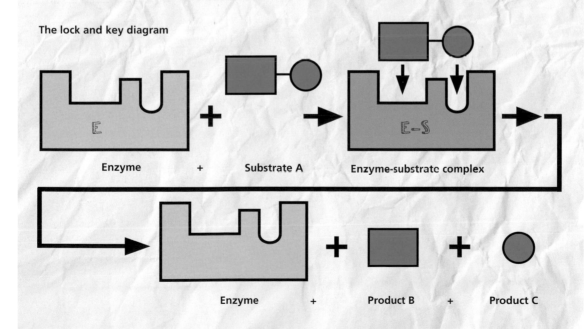

| Enzyme | + | Substrate A | Enzyme-substrate complex |

| Enzyme | + | Product B | + | Product C |

Types of Cell

The body contains many distinct kinds of cells, each with a unique structure that is specialized to perform particular functions.

Different types of cell perform specific functions in the body. For example, specialized white blood cells in your body have an abundance of ribosomes because they need to produce lots of protective proteins called antibodies. Skeletal muscle cells have an abundance of mitochondria because they require much more energy than other cells for the contraction of muscles in maintaining posture and movement. Digestive cells and endocrine (glandular) cells have lots of ribosomes to produce digestive enzymes and hormones respectively. They also have plenty of golgi bodies that package these digestive enzymes and hormones so they can be released or secreted into the body. (Digestive and endocrine cells are "secretory" cells, which means they release chemicals into the body.)

Monitoring Movement

To maintain their specialized functions, cells need to monitor the chemical environment inside and outside the cells of the body. Receptors within the cell or on its outer membrane detect any changes. These receptors inform genes, which produce (start) or inhibit (stop) certain actions.

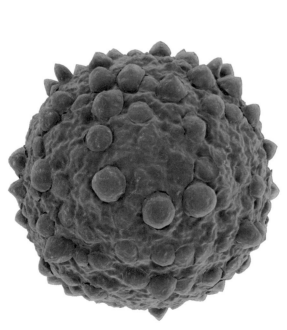

White blood cells protect the body from infectious diseases and foreign substances.

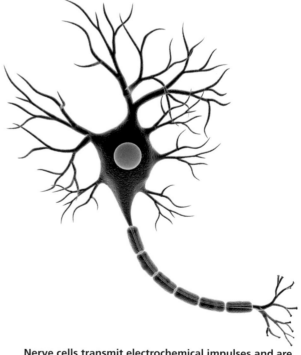

Nerve cells transmit electrochemical impulses and are the basic units of the body's nervous system.

Cells, Tissues and Body Structure

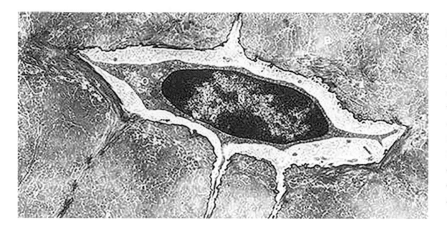

Bone cells (osteocytes, left) are found in fully formed bone tissue. They are long-lived and can transmit signals to other osteocytes if any part of the bone becomes deformed through muscular activity. Premature cell death or dysfunction of osteocytes can lead to diseases such as osteoporosis.

Body and Cell Secrets

- The body is composed of trillions of microscopic cells, with an average diameter of 0.02 mm (1/2500 in).
- The largest cell in the body is the female egg, which is approximately 0.5 mm (1/50 in) diameter and just visible to the naked eye.
- The longest cells are the nerve cells supplying your toes, measuring as much as 1.2 m (3.9 ft) in length, but even these are microscopically thin, meaning we need a microscope to see them!
- Five million of your body cells die every second, but most are renewed.

- White blood cells are used in the fight against infection and may survive for only a few hours.
- Your gut lining cells live for approximately three days.
- Red blood cells live for an average 120 days.
- Bone cells live for about 20 years.
- Your brain cells cannot regenerate, meaning they must last a lifetime, as once they have died, they are gone forever.

Red blood cells absorb and transport oxygen, and deliver it to the body's tissues.

Female eggs (ova) are reproductive cells that grow in the ovaries.

Cell Division

Cell division is an important "basic need of life" as it enables our bodies to survive, grow and repair, and ensures that genetic material is passed on to the next generation.

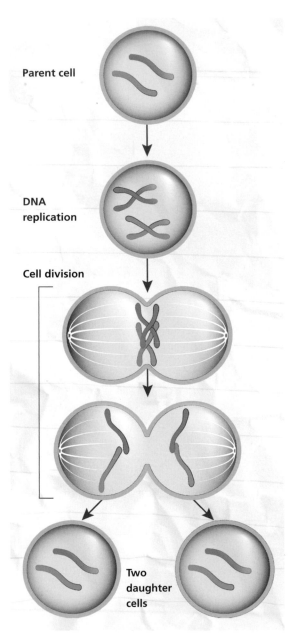

Parent cell

DNA replication

Cell division

Two daughter cells

Cell division makes sure that the body's genetic material is transmitted from one cell to another, and from one generation to another. It also enables cell growth and specialization, and growth in specific stages of human development, such as from baby to infant and infant to young child. In addition, it ensures that dying, diseased, worn-out and damaged cells are replaced to maintain the structural and functional integrity of the human body.

There are two types of cell division – body cell division called mitosis and reproductive cell division called meiosis. In each case, the dividing parent cells produce daughter cells. (Sorry, boys, it's nothing personal!)

Keeping Up the Chromosomes

Mitosis, or duplication division, ensures that the daughter cells have the same number of chromosomes – and therefore the same number of genes and DNA, as chromosomes are made up of genes, which in turn are made up of DNA – as the parent cell. For this to happen, the 23 pairs of chromosomes of the parent cell must first be duplicated so that one copy (23 pairs) can be passed into each daughter cell.

During mitosis, the DNA of the parent cell duplicate and form X-shaped chromosomes. After this, the cell divides and the two daughter cells finally separate, each of them an identical copy of the parent cell, with exactly the same genetic information.

Making Sex Cells

Reproductive cell division, called meiosis, reduces the genetic material of two cells to produce male and female sex cells.

Meiosis ("reduction division") occurs in the male and female sex organs (gonads) during the production of the male and female sex cells (spermatozoa and ova, respectively). Meiosis ensures that the daughter cells have half the number of chromosomes of a normal cell (23 chromosomes, half of the 46 chromosomes of the parent cell). This reduction division is necessary so that the normal chromosomal number is restored when two sex cells fuse during fertilization. The new cell (zygote) then divides by mitosis (*see* previous page) and ultimately produces trillions of specialized cells that make up the specific tissues and organ systems of the human body. The joining together of parental genetic information at fertilization is the reason why we can see similarities in looks and behaviour to both our parents.

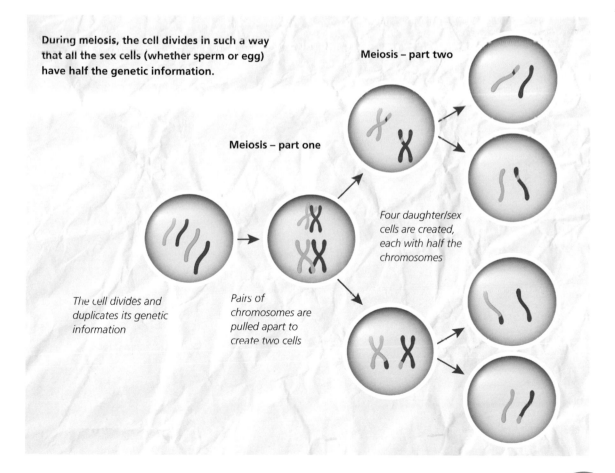

During meiosis, the cell divides in such a way that all the sex cells (whether sperm or egg) have half the genetic information.

Meiosis – part one

Meiosis – part two

The cell divides and duplicates its genetic information

Pairs of chromosomes are pulled apart to create two cells

Four daughter/sex cells are created, each with half the chromosomes

Cell Specialization

During the development of an unborn child, every cell starts to differentiate into specialized cells so they can perform different functions within the body. For example, the movement of the body's skeleton is supplied by the specialized skeletal muscle cells. Similarly, the generation of electrochemical impulses is provided by special nerve cells, the fighting of infection is carried out by white blood cells, and the transportation of oxygen is performed by red blood cells.

The process that transforms the original unspecialized cells into specialized ones is called cell differentiation and specialization, a process determined by the switching "on" or "off" of certain genes within the cell. There are over 200 different types of cell within the body. However, it's not fully understood how and why the chemical signals control the different cells and programme them for specific functions in a particular location of the body.

The switching on or off of certain genes dictates the form and function of the cell as it matures.

Chromosomal Abnormalities

A typical cell divides every 25 hours, although some specialized cells (such as most nerve cells) may not divide again. DNA duplication should conserve the genetic make-up of a cell, but errors can occur during mitosis, which can lead to chromosomal abnormalities in daughter cells. Depending on the extent of the abnormality, errors during cell division may not affect the function of an adult body as only one of many thousands of cells may be affected. With age, however, an accumulation of errors could contribute to declining function in the body or even to the development of certain diseases, such as cancers. When cells are specializing and growing in the unborn child, errors in cell division can have a pronounced influence on tissue development and function. An extra chromosome in the fertilized egg (zygote) is called a trisomy and this can have a visible effect on the development of a human. Cells derived from the zygote will have an extra chromosome (and extra genes), and will result in certain signs and symptoms in the body. The best known example of a genetic disorder caused by a trisomy is trisomy 21, also known as Down's syndrome.

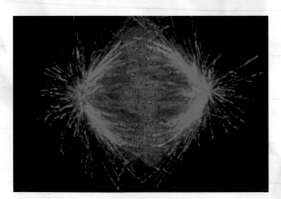

This image shows the moment when human chromosomes (blue) duplicate and divide during mitosis.

Tissues

The tissues in the body perform a variety of functions, from cushioning delicate organs and lining body surfaces, to expanding and contracting to produce movement in the body.

As cells divide by mitosis and move to new locations in the body, their surrounding chemistry changes. These differing chemicals may be responsible for the switching on and off of specific genes, causing cell differentiation and specialization, and the consequential joining together of these specialized cells with similar properties to form tissues. The 200 different kinds of cells fall into four main categories and make up four types of tissue. Each of these cell types can be further divided into subtypes to perform specialized functions.

Epithelial Tissue

These tissues are found in many areas of the body. They line body surfaces, such as the skin, and cavities, such as the stomach, where they have a protective function. Your epithelia tissues may be simple (one cell thick) or compound (more than one cell thick). Some epithelial tissues, known as glandular epithelia, produce bodily secretions, such as tears to lubricate the eyes and sweat to regulate body temperature. The tissue stops the cavity from drying out and also keeps out substances that could pose a threat, such as pathogens (microorganisms that cause disease) – or even sunlight!

Squamous epithelium is a layer of single cells that is usually found where small molecules need to pass from one place to another, such as alveoli in the lungs and blood capillaries.

Epithelial tissue consists of sheets of cells held together by a basement membrane. For example, the lining of the human bladder is made of compound epithelial tissue. Its wrinkled surface allows the bladder to expand and contract as it fills with and empties out urine.

Muscular Tissue

Your muscles are a special type of tissue, which contract to produce movement. There are three different types of muscle tissue in the body:

- **voluntary** (also called skeletal or striated muscle)
- **involuntary** (also known as unstriated or smooth muscle)
- **cardiac** (heart muscle).

Muscle cells are often referred to as muscle fibres, because they are long and cylindrical. These elongated cells or fibres can range from several millimetres ($^1/_3$inch) to about 10 centimetres (4 inches) in length.

Muscles contract because they have two types of protein fibres: actin and myosin, which slide over each other to shorten the muscle. Skeletal muscle makes up the flesh of limbs and the body's torso and it moves the skeleton. When stimulated by nerve fibres, these muscles can perform rapid, powerful contractions, but they get tired very quickly and use up their energy supplies faster than non-skeletal muscle. As a result, skeletal muscles have many mitochondria (*see* page 12) and require a good blood supply to bring oxygen and nutrients to quickly replenish energy supplies.

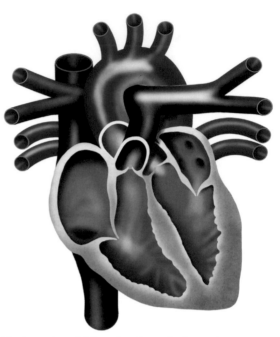

The heart has thick walls of muscle tissue that contract continually throughout a person's lifetime.

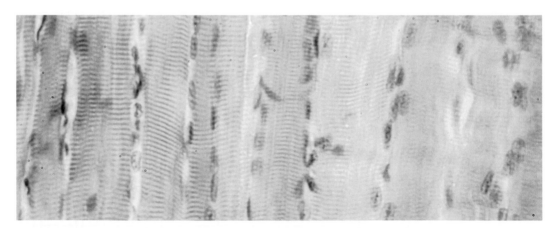

The striated or striped appearance of skeletal muscle (which gives this type of muscle its alternative name) is formed by the orientation of the muscle fibres.

Cells, Tissues and Body Structure

protect against pathogens that can cause disease, and last, but not least, storing energy.

There are many types of connective tissue, including white fibrous tissue, adipose tissue, loose areola tissue and yellow elastic connective tissue. It is also made up of different types of cells, such as bone cells, cartilage cells and blood cells. Despite these variations, all types of connective tissue consist of the same three components:

- **a ground substance** called a matrix, which may be fluid, jelly-like or solid
- **cells** that produce the matrix
- **protein fibres** made up of collagen, elastin or reticulin.

Sensory nerves carry signals to the spinal cord and/or the brain from receptors around the body, while motor nerves carry instructions from the brain and spinal cord to different body parts.

Nervous Tissue

Nervous tissue is made up of nerve cells, also known as neurons, which are specialized to generate and conduct electrochemical impulses. There are four main types of neurons: brain cells; sensory neurons, which carry impulses from sensory receptors to the spinal cord and/or the brain; motor neurons, which carry instructions from the brain and/or spinal cord to the tissues of the body; and interneurons, which connect sensory and motor neurons together within the central nervous system (the brain and spinal cord).

Connective Tissue

Connective tissue is the most common type of tissue in the body. It performs many functions, from providing a structural framework for your body, supporting and binding different tissues within organs, and surrounding and cushioning delicate organs, to transporting secretions and fluids from one region to another. They also

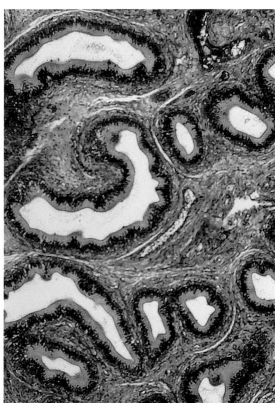

In this cross-section of a windpipe (trachea), the blue highlighted area is a layer of areola connective tissue, which joins together all the layers, including one that contains blood vessels (shown in red).

Organs and Systems

An organ is a group of different tissues working together to perform a specific function. Organs that work together form organ systems.

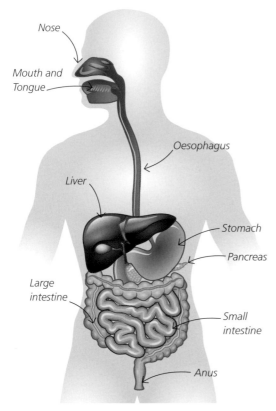

Nose

Mouth and Tongue

Oesophagus

Liver

Stomach

Pancreas

Large intestine

Small intestine

Anus

The digestive system breaks down, absorbs and removes food and is made up of several organs.

Each of your organs performs a specific activity that is vital for the survival of the body. Examples of organs are the heart, stomach, liver, brain and skin. Most organs contain all four of the different types of tissues (connective, nervous, muscular and epithelial). In the stomach, for example, the inside epithelial lining secretes gastric juice and absorbs chemicals like alcohol and glucose. The wall of the stomach, however, also contains muscle tissue to aid contraction of the stomach and to help with the breakdown of food, as well as nervous tissue to help regulate the stomach and connective tissue to bind the other tissues together.

Every organ forms a part of a larger system that may contain more than one organ. As an example, the heart is part of the cardiovascular system. An organ system is a group of organs that act together to perform a specific bodily function. The respiratory system maintains the levels of oxygen and carbon dioxide in the blood. These systems work with each other in a coordinated way to maintain the functions of the body. Each level of organization (cell, tissue and organ system) is instrumental in sustaining the basic needs of the human body.

Body Fluids

Cells in the body are bathed in interstitial (tissue) fluid, which is derived from blood plasma. Tissue fluid supplies the cells with oxygen and nutrients, carries away cell products, such as hormones, digestive enzymes and tears, and flushes away cell wastes, such as carbon dioxide.

- **The amount of tissue fluid bathing your cells is about 11 litres – that's nearly 20 pints of milk!**
- **The fluid outside your cells (extracellular fluid) is made up of tissue fluid and 3 litres (5 pints) of blood plasma.**
- **The amount of fluid inside your cells (intracellular fluid) totals a massive 28 litres – nearly 50 pints of milk.**

Cells, Tissues and Body Structure

Cytology in Medicine

Cytology includes the branches of biology and medicine concerned with the study of cells. Its findings help medical professionals better understand ill health and how to treat it more effectively.

Understanding the structure, function and the needs of human cells can also help us to understand how tissue and organ abnormalities lead to ill health. As an example, the condition Obstructive Respiratory Disease causes low supplies of oxygen and high carbon dioxide levels in the blood, which can result in cell death (necrosis) in the patient. The body attempts to correct these imbalances after its receptors detect the problem and send information to the respiratory control centre in the brain.

Cytology is concerned with the study of cells and medicine is concerned with the science of the body when it goes wrong and its findings help medical professionals better understand ill health and how to treat it more effectively. Having analyzed the imbalances in the blood, the brain then sends information to the respiratory muscles, which attempt to reverse the problem by increasing the rate and depth of breathing. If this fails to establish equilibrium (homeostasis), then medical intervention may be necessary as the body is unable to resolve the problem by itself.

Diagnosis

Diagnosis and treatment, however, are often more complex. As different parts of the body depend on each other, a disorder that arises at a cellular level and leads to a failure of one functional group of cells can often lead to a deterioration of other groups of cells and different parts of the body. This is shown in the diverse signs and symptoms that can arise from ill health, all of which require medical intervention. For example, a patient who has had a long-standing (chronic) Obstructive Respiratory Disease may display signs and symptoms that reflect poor functioning of not only the heart, but also the kidneys.

This is why you should never try to diagnose yourself as sometimes even doctors find it tricky and they have spent many years dedicated to studying medicine and the body.

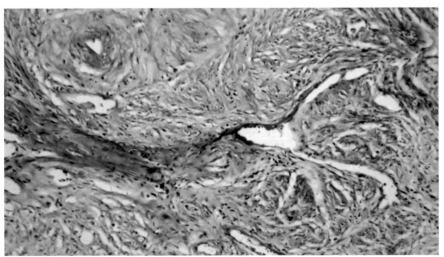

Cytology can be used by medics to diagnose conditions in patients. In this instance, a tissue sample from the cervix is showing development of cancer cells.

Benign and Malignant Cells

A tumour is a swelling produced by an abnormal accelerated period of growth and reproduction of cells. A tumour is classified as benign or malignant. Benign tumours are non-cancerous, do not spread to other parts of the body and are usually deemed safe. Malignant tumours are cancerous, spread to other areas and infiltrate other tissues and organs. This spread (metastasis) is dangerous and difficult to control. Malignant tumours can be treated by surgically removing the tumours or killing the cancer cells using radiotherapy (deep X-ray radiation). Heating and freezing can also be particularly effective prior to metastasis. Radiotherapy prevents cell division by breaking down chemicals in the chromosome and the cytoplasm. Chemotherapy, another treatment involving the use of chemicals, kills cancerous tissue and stops the reproduction of cancerous cells by inhibiting the separation of duplicated chromosomes.

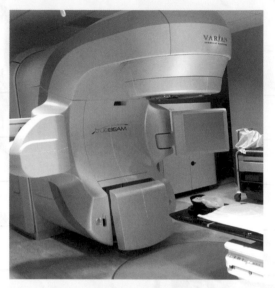

A linear accelerator used during high-precision radiotherapy to treat various cancers.

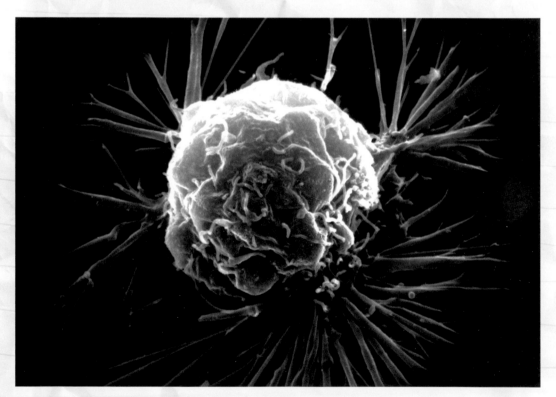

A cancerous cell, such as this malignant breast cell, will spread to other tissues and organs.

Cells, Tissues and Body Structure

Chapter 2
THE SKIN, SKELETON AND MUSCLES

The Skin

How you look is dictated by the skin and the structures associated with it, such as hair, nails and eyelashes. The body also has a very definite shape, which is largely decided by two body systems – the skeletal system and the muscular system.

Covering It All

The skin is your body's largest organ. It covers the outer surface of the body and extends into the oral and anal canals. Being the main boundary between the body and the outside world, it's not surprising that the skin is the first line of defence against harmful organisms, or pathogens. It also helps to regulate your body temperature and hydration levels, and provides sensory information about your surroundings, such as sensitivity to touch, heat and cold. Being the most visible part of your body, the skin also makes clear the effects of ageing, as seen when it wrinkles with age.

Your skin has two main layers – the outer epidermis and the inner dermis. Beneath these layers is the subcutaneous (fat) layer, which is sometimes called the superficial fascia, since it also includes part of the connective tissue that covers muscles. The layered structure of the skin provides a physical barrier to trauma, while the skin secretions also provide a degree of chemical protection against bacteria. The impervious (waterproof) nature of skin, and its role as a physical barrier, also protects the body from chemical agents like bacterial toxins, which may make the body ill.

The top layer of skin, the epidermis, is made up of flattened, dead skin cells. At the bottom of these is the living epidermal layer where new skin cells are created before rising to the surface. Beneath the epidermis is the dermis, which contains most of the sensory receptors, as well as the hair follicles and sweat glands. Note the bumps, or dermis projections, known as papillae, which stick up from the dermis into the epidermis. Many of these have tiny blood vessels to carry nutrients up to the epidermis, since the upper layer does not have its own blood supply.

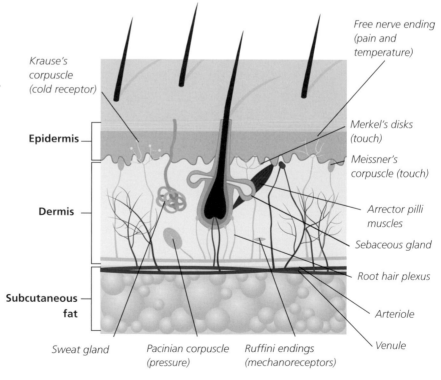

Krause's corpuscle (cold receptor)

Epidermis

Dermis

Subcutaneous fat

Free nerve ending (pain and temperature)

Merkel's disks (touch)

Meissner's corpuscle (touch)

Arrector pilli muscles

Sebaceous gland

Root hair plexus

Arteriole

Venule

Sweat gland

Pacinian corpuscle (pressure)

Ruffini endings (mechanoreceptors)

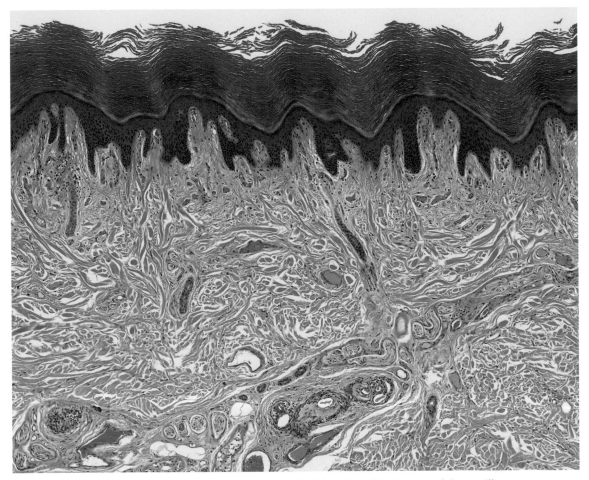

This micrograph of the skin clearly shows the dermis and epidermis and the bumps of the papillae.

Epidermis

As the epidermis is your body's first physical barrier to the outside world, it needs to be tough to provide sufficient protection. The epidermis has multiple layers, the lowest layer – the basal layer – is a single layer of cells connected to a basement ("cementing") membrane, which separates the epidermis from the dermis. The cells in this basal layer continually divide and the daughter cells produce all the outer layers within the epidermis. This is why the basal layer is often referred to as the "germinal" layer. Dotted around these lower layers are the melanocytes. These special cells produce the protein pigment melanin that gives skin its colour.

As the cells ascend, each one of them gradually loses its nucleus and becomes filled with a protein called keratin, which makes the epidermis tough and waterproof. As a result, the cells become flat, hard and die, creating a cornified outer layer. This outermost layer gives the epidermis the toughness needed to act as a barrier against external physical stresses and other threats, such as bacteria and harmful chemicals. In adults, the epidermis is 0.5 to 3 mm ($^1/_{50}$ to $^1/_{10}$ in) thick, depending upon the thickness of the cornified layer and the physical stresses placed on that area of skin. For instance, the epidermis of the eyelids is very thin, while the epidermis of the soles of the feet is very thick. As the outer layer of skin, it is not surprising that cells

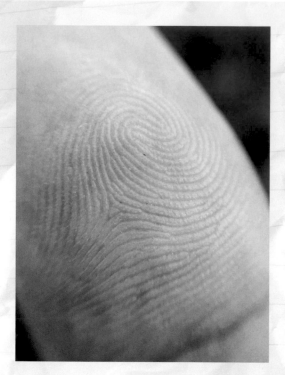

are continuously lost from the epidermis during day-to-day living through wear and tear. The attrition (loss) is substantial – in fact most of the "house dust" in bedding is made up of these lost cells!

The epidermis needs to be replaced every 35 to 45 days and the demands for this cell replacement need to be met by a network of blood vessels within the dermis, since the epidermis is "avascular", meaning it doesn't have its own blood supply.

Dermis

The dermis is usually thicker than the epidermis. It is composed largely of connective tissue, containing collagen that gives your skin strength, and elastin fibres that give your skin its elasticity. The spaces between the fibres contain blood and lymphatic vessels, nerves, sensory receptors (responding to touch, pressure, temperature and pain), hair follicles (though these are epidermal in origin), the hair's arrector muscles (which, when contracted,

cause the hair to stand upright) and ducts of subcutaneous glands (the glands themselves lie in the subcutaneous layer).

The upper region of the dermis has small bumps called dermal papillae. These project into the epidermis and produce your unique "fingerprints". The papillae contain touch-sensitive receptors (Meissner's corpuscles) and loops of blood capillaries. These loops are arranged vertically, and are visible as small pinpricks of blood when the epidermis is grazed.

Hair

Hairs are found on almost every part of your body, and while you might have the ones on your head cut or styled regularly, they carry out a wide range of functions.

Hair is made of the waterproof protein keratin and dead cells, and is produced by germinal or basal cells of the epidermis. These cells are more active than the other cells of the epidermis, so they need a better blood supply. It therefore makes sense that the follicles and nail beds extend deeper into the skin where blood flow is greater.

Tough and Waterproof

One function of hair is to protect underlying structures. For example, head hair provides some shade from the sun and it also reduces heat loss on a cold day. Some body parts, such as the lips, have no hair, while other places have very little, and this is thought to reduce insulation and aid heat loss. Hair can also be stubbly in places, such as the ears and nose, helping to filter air and make sure harmful microbes do not enter the body. However, even where there is very little hair, there are touch receptors nearby, giving the hairs a sensory role.

Structurally, a hair has three parts – the root, the follicle and the shaft. During active growth, the hair root within each follicle is surrounded by a live tissue, called the hair bulb. This bulb contains a layer of dividing basal (germinal) cells, and, as these new cells form, older ones die and are pushed up to form the root and shaft of the hair. Approximately 70 to 100 hairs are normally removed from the scalp each day and will be replaced by new hairs (so it's not just your fluffy pets that moult!). However, if you possess the genes for baldness, new hairs will not always grow to replace the lost ones.

The number and density of hairs varies depending on the location on the body. It is usually thickest on the scalp and should be non-existent on the palms of the hands and soles of the feet.

As the arrector muscles pull on the base of the hair to raise it, they create a small bump in the skin surface, producing the appearance of goose bumps.

In response to a possible threat, this cat has raised the hairs in its coat in an attempt to make itself look bigger and more imposing.

Glands and Muscles

Next to the follicles are the sebaceous glands and small arrector muscles. Sebaceous glands are found all over the body except the eyelids (and no one knows for sure why they are absent from there). They secrete sebum, an oily substance containing fats. This helps to preserve the suppleness (softness) of hair and skin, and to waterproof the epidermis. Sebum, along with sweat, also helps to protect the body – a mixture of sweat and sebum is responsible for an acid pH on the skin surface, and this discourages the establishment of alkali-liking microbes which can cause potential infection and illness. The arrector muscles, when contracted, cause the hair to stand on end, a process called piloerection. In animals, this is an important temperature-controlling mechanism, trapping air close to the body to increase insulation. In humans, this produces the "goose bump" flesh appearance, but this is not considered an important temperature-controlling mechanism in humans. Piloerection is also seen during moments of fear – the hairs on some animals, such as monkeys, dogs and cats, stand up on the back of their necks. This is a threatening or defence response called the "fight or flight response", which helps us decide what to do when we feel a certain situation is dangerous. Perhaps, this is why people claim that their hair standing on end is a sign of a ghost in the room!

The Skin, Skeleton and Muscles

Sweat Glands

Sweat glands are long, coiled hollow tubes. Sweat, or perspiration, is produced in the coiled parts deep in the dermis and this sweat is transported by the sweat duct up to the surface.

Types of Sweat Gland

You have two main types of sweat gland:

1. Apocrine sweat glands are found on hair-bearing skin such as your armpits, scalp and groin. They become active at puberty and secrete sweat, containing proteins, fats and sugars, into hair follicles rather than onto the skin's surface. This type of sweat is broken down by skin bacteria to produce an unpleasant body odour. Apocrine sweat also contains pheromones – chemicals that are associated with sexual attraction. Antiperspirants prevent the release of such chemical messengers, which might seem detrimental to sexual attraction, but they also prevent bacterial growth and body odour, and so at least prevent the opposite sex being deterred by your smelly pits when you've skipped a shower or two!

2. Eccrine sweat glands are distributed over nearly all of the body, with the greatest density being on the palms and soles, although, they are absent from the lip margins, penis, labia minora and outer ear. The main role of eccrine sweat is in temperature regulation since the evaporation of sweat from the body surface has cooling properties. Eccrine sweat contains water, salt and other waste products. Since water is a major part of sweat, it is important to change your water intake depending upon environmental temperatures. In cool temperate conditions, sweat accounts for almost 10 per cent of the water output from the body. In very hot weather, sweat secretion may rise to as much as 4 litres (7 pints) per day and so exceed the total excreted by other routes, including in urine, meaning it is important to keep hydrated in hot weather to make up for the water lost in excreted sweat.

Ceruminous glands are modified sweat glands found within the skin of the auditory canal. Their secretions are mixed with sebum from sebaceous glands in the dermis to form a sticky, wax-like substance called cerumen, or earwax. This provides a barrier to particulate matter, and so protects the eardrum (tympanic membrane). Occasionally, the secretion of cerumen may be excessive and induce conditions that promote troublesome bacterial

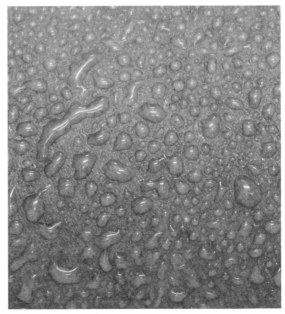

Watery sweat takes heat away from the body as it evaporates, cooling you down.

growth, which may lead to ear, nose and throat infections. Hearing will also be impeded if excessive wax is present, which is why it is good to make sure your ears are cleaned regularly.

Nails

Nails strengthen the tips of your fingers and toes meaning they protect them from damage and improve your grip on small objects. Nails are made up of keratin, just like your hair! Each nail consists of three parts: the root, the nail plate and the free edge at the tip. The lunula is the visible crescent-shaped part of the nail root that produces new nails. Lunula is pronounced on the thumbnail, but barely visible on the little finger. Over the lunula, a thin fold of cuticle is usually pulled up from the enclosing skin as the nail grows. Nail growth begins from the root to the tip of the finger and will keep growing until you cut them (just see the Guinness Book of World Records if you want proof!).

Subcutaneous Layer and Glands

This layer consists of connective tissue and adipose (fat) tissue attaches the dermis to the underlying tissues. The presence of relatively few collagen fibres means that the subcutaneous layer allows flexibility of movement of skin across the underlying tissue, which helps to prevent shearing injuries due to friction.

Removing Excess Earwax

If earwax becomes excessive, the person may express a "muffling" of hearing in the ear, or even soreness, as the tympanic membrane becomes inflamed secondary to infection behind the wax. The ears need to be syringed to flush out the wax (below), possibly after a period of several days of softening the wax with a solution applied by a dropper.

The adipose tissue is useful because the subcutaneous layer is readily stretched, so it can store substantial amounts of fat, and provides a layer of insulation to reduce heat transfer from the body to the environment. From a clinical perspective, the presence of loose connective tissue provides an ideal site for injections, since the volume of liquid to be injected is accommodated more easily and is therefore less painful.

Lee Redmond of the USA started growing her nails in 1979 and, by 2008, they had reached a total length of 8.65 metres (28.4 feet).

The Skin, Skeleton and Muscles

The Skeleton

An adult skeleton is made up of 206 different bones that help to keep you in shape and move you around from place to place.

A common description from people who view the human skeleton in isolation is that it is a collection of oddly-shaped, dry and dead bones. While there is no doubt that the bones are oddly-shaped, the reality is that a skeleton is made up of a dynamic, living connective tissue that is continually shaped and re-shaped throughout a person's life. Bones shape your body and, with the aid of the skeletal muscles, allow your body to move. Parts of the skeleton protect your vital organs from impact injuries. For example, the bony "box" of the skull protects the brain, and the cage-like structure of the ribs protects the heart and lungs. Found inside many of your bones, bone marrow is the main site of blood cell production, and bone tissue also stores and supplies the body with key materials, such as calcium and phosphate. Other tissues found in bones include cartilage, nerves and blood and lymphatic vessels, and all of these play very important parts in keeping the body moving and working properly.

Adult vs Child

The adult skeleton contains 206 bones (at birth, babies have more than 300, some of which fuse as they grow), and these can be divided into the axial and appendicular skeletons. The axial skeleton forms the vertical axis of the body, and is made up of the skull, vertebral column (backbone), ribs and sternum (breastbone). The appendicular skeleton is so named because these bones are "appended" (joined) onto the central axial skeleton. It is made up of the limb bones found in the arms and legs, and the bones of the limb girdles, including the pelvis and the shoulder blades, or scapulae.

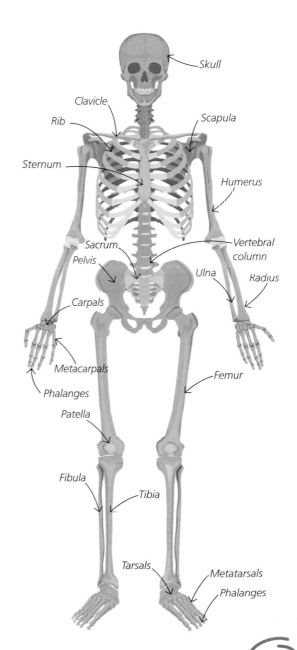

Skull
Clavicle
Rib
Scapula
Sternum
Humerus
Sacrum
Vertebral column
Pelvis
Ulna
Radius
Carpals
Metacarpals
Femur
Phalanges
Patella
Fibula
Tibia
Tarsals
Metatarsals
Phalanges

Bones Shapes

Bones come in various shapes and sizes depending on their activity.

Short Bones

These include the wrist and the ankle. They are not as strong as long bones, but when collected together produce great flexibility.

Flat Bones

These are thin, plate-like bones, such as the skull bones, and provide protection for underlying structures. Or they provide a large area for the attachment of large muscles, for example the shoulder blade (scapula).

Long Bones

These include the limb bones, and they provide a wide scope for body movement and also absorb the stresses of body weight.

Sesamoid Bones

The kneecap (patella) is the main example of this type of bone and it has a tendon attached to make the knee joint much more stable.

Irregular Bones

These have complex shapes that relate to their activity. For example, the backbone (vertebral column) has multiple sites providing a large surface area for the attachment of the back muscles and ligaments. Together, they provide great flexibility.

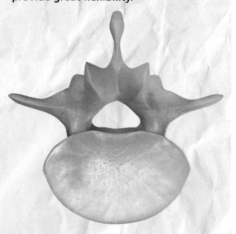

Shapes and Structure

Bones have many regular external features. These are determined by where muscles and ligaments need to attach, how joints with other bones are shaped, and sometimes for the passage of blood and lymphatic vessels and nerves that run through the bone. Bone is a rigid type of connective-tissue that makes up the human skeleton. There are two types of bone tissue: compact (cortical) and spongy (cancellous) bone, and they get their names from their appearance (more of that later). Inside this tissue, bone cells, called osteocytes, are surrounded by a tough matrix of collagen fibres and protein-bone fibres called ossein, which have been strengthened by calcium and phosphate salts. This matrix gives bones their characteristic strength and rigidity.

The Skin, Skeleton and Muscles

Long Bones

Each long bone is made up of a shaft and two ends. The shaft consists of a thick outer layer of compact bone surrounding spongy bone, within which is a central cavity containing bone marrow. The end of the bone consists mainly of spongy bone covered by a thin outer layer of compact bone. Near the end of long bones is the epiphyseal plate, which under the influence of growth hormone (from the pituitary gland), lengthens during periods of growth. Too much growth hormone produces a condition called "pituitary giantism", while too little causes a condition called "pituitary dwarfism". Bone tissue can grow very quickly in the first two decades of life. After this, the epiphyseal plate fuses, or becomes ossified, and as such, they no longer respond to growth hormones, so growth in bone length stops and this is when we stop getting taller! However, your bones continue to repair and regenerate themselves throughout life at sites other than the epiphyseal plates.

This image (below) shows a cross-section through a long bone, in this case a femur. The whole bone is covered with a layer called the periosteum. Beneath this is the solid compact bone. Spongy bone is the meshed honeycomb structure below compact bone. This mesh is filled with bone marrow where blood cells are made.

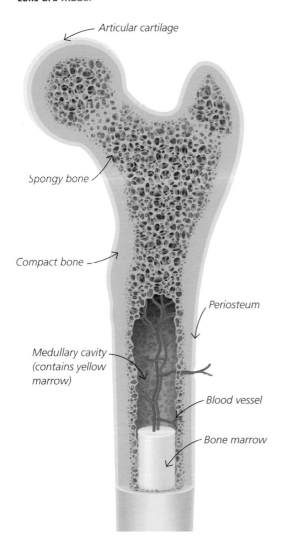

Articular cartilage

Spongy bone

Compact bone

Periosteum

Medullary cavity (contains yellow marrow)

Blood vessel

Bone marrow

Bone Secrets

- There are 26 bones in the human foot, and 27 bones (including the wrist) in the human hand.
- The femur (thighbone) is the strongest and longest bone, whereas the your stapes, in the middle ear, is the lightest and smallest in your skeleton.
- Your arm bones are the most commonly broken bone in adulthood, whereas the collarbone is the most commonly broken bone in childhood.
- Compact bone is the second hardest material in the human body after your tooth enamel.
- 80 per cent of the body's bone weight comes from the compact bone. The other 20 per cent comes from spongy bone, even though spongy bone has a surface area 10 times greater than compact bone.
- Bone is five times stronger than a steel bar of the same weight.
- Bone marrow is found within the spongy bone of your long bones. Red bone marrow produces red blood cells, while yellow marrow stores fat.

Demineralizing Bone

Osteoporosis is when the bone retains its structure, but it loses a lot of its bone mass and therefore its strength. The loss of bone mineral is normal during a lifetime, and usually starts when a person reaches their forties.

Accompanying this loss of strength and bone mass is a reduction is the skeleton's ability to support the body, and this can cause an increase in the risk of fractures, particularly of the neck or femur, and consequential loss of mobility.

Osteoporosis is common in the elderly and more frequent in women, since the hormonal changes associated with menopause increase the rate of mineral loss. Osteomalacia, usually caused by a deficiency of vitamin D, is a rare condition in Western countries. Vitamin D promotes calcium uptake from the bowel and helps to maintain the levels of calcium in blood plasma. Low levels of this vitamin mean that less calcium is available for bone materialization. It differs from osteoporosis in that the bone maintains a relatively balanced protein matrix but the ratio of mineral to protein declines, making the bones softer and thus less supportive.

This deficiency can be due to an inadequate dietary intake of the vitamin, or a lack of its synthesized source due to the lack of exposure of the body to sunlight. It could also occur as a result of a malabsorption of the vitamin through a failure of the liver to activate it. The problem may also be caused by specific tumours. These tumours can secrete the hormone parathyroid and so promote excessive calcium resorption from bone.

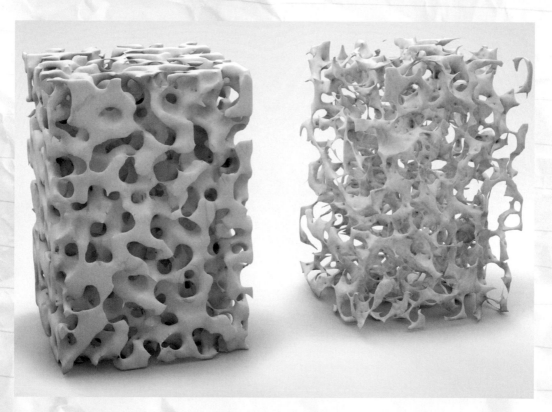

This graphic shows models of normal bone tissue (left) and bone tissue with osteoporosis (right).

Joints

Strictly speaking, a joint is simply where two bones or cartilage and bone meet. While many joints in the body allow for movement, there are a few where the bones are held rigidly together.

Joints are divided into:

- **Fixed joints** – those between bones, which provide no movement.
- **Cartilaginous joints** – those between bone and cartilage that allow slight movement.
- **Synovial joints** – those that allow movement and are called "mobile" joints.

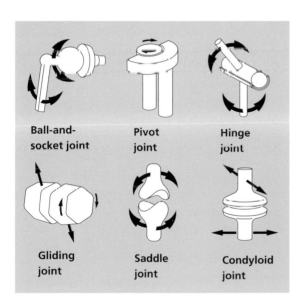

Ball-and-socket joint

Pivot joint

Hinge joint

Gliding joint

Saddle joint

Condyloid joint

Arthritis

Inflammatory conditions of synovial joints are collectively known as arthritis. Osteoarthritis is a degenerative condition where there is a breakdown of synovial cartilage resulting in the flaking and cracking and eventual loss of joint cartilage. As the cartilage gets thinner, there is friction between the two bones, and this causes bone erosion, inflammatory responses such as swelling (oedema) and pain in the affected joint. Over time, calcification of the joint capsule and production of bony outgrowths, known as osteophytes, may occur resulting in reduced joint flexibility. Osteoarthritis can be caused by excessive stresses and trauma to the joint over time, as "wear and tear" take their toll, or it may be a genetic condition that has been inherited.

Rheumatoid arthritis (RA) is an autoimmune disorder of connective tissue of the joint capsule, articular cartilage and ligaments. The affected joints become deformed, swollen and painful. The cause of the RA is obscure, and there may be a several factors involved, such as genetic susceptibility and hormonal inbalances.

Joints in the Foetal Skull

At birth, a child's head size is large compared to the rest of the body and the brain weighs about 25 per cent of its final adult weight. The brain grows rapidly and this is helped by the incomplete joints of the eight bones that make up the skull. There are spaces, called fontanelles, between the infant bones that are covered by fibrous membranes. Most fontanelles close during the first few months, but the anterior fontanelle is not fully closed for 18 to 24 months. The actual joints (sutures) remain flexible until you are about 5 years old, by which time growth of your brain and cranium has slowed considerably.

Skeletal Muscle

Quite simply, without skeletal muscles, you would not be able to move! These muscles are attached to the bones of your skeleton and are mostly under your voluntary control. The majority of skeletal muscle lies just under the skin.

When you look at fully-labelled diagrams of the body muscles, it can be baffling, as you are confronted with many muscle names, such as the flexor digitorum superficialis. As with many of the anatomical features of the human body, the naming of muscles relates to various features of the muscle, for example:

- **size**, for example the pectoral major, the large (major) muscle of the chest (pectoral)
- **location**, for example the tibialis anterior, lies in front (anterior) of the shin bone (tibia)

- **shape**, for example the deltoid muscle of the shoulder (delta – "triangle-shaped")
- **number of "heads" of muscle**, for example the triceps muscle (tri – three heads)
- **movement type**, for example adductors indicate muscle moves towards the midline of the body.

The situation is even more complex than this because frequently, the naming of muscles relates to one or more features. For example, the levator (lifter) palpebrae (lip) superioris (upper), where the name indicates both its position and action.

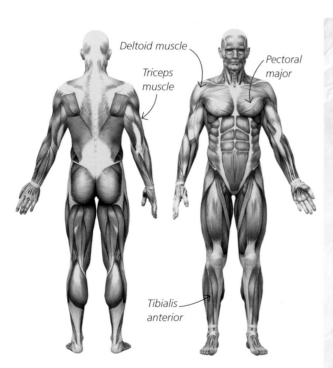

Deltoid muscle

Triceps muscle

Pectoral major

Tibialis anterior

Human skeletal muscles: back and front views

Secrets About the Muscles

- There are over 600 muscles in the body and they make up between 30 to 50 per cent of your bodyweight.

- Most of your skeletal muscles have "mirror images" on the other side of the body.

- Some of your skeletal muscles are always slightly contracted to offset the effects of gravity.

- The largest single tendon in the body is the Achilles' tendon, named after the Greek hero, which attaches the large calf muscle (the gastrocnemius) to the heel bone.

- The smallest tendon, just over a millimetre in length, is the stapedius, which stabilizes the smallest bone in the body, the stapes in the middle ear.

Muscle microstructure – sheaths within a muscle

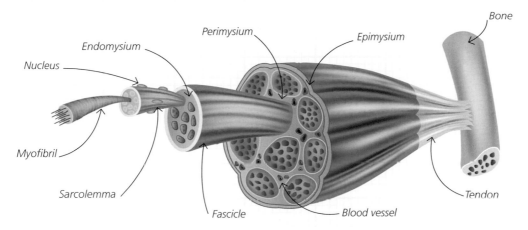

Muscle Architecture

As a rule, each muscle has a wide central region (belly) and two ends. These ends are tough cords of connective tissue, called tendons, that attach to either bone or cartilage.

Muscles are covered in layers of dense, connective tissue (the deep fascia), which acts to separate them so they can act independently. Below this fascia lie three sheaths of connective tissue that provide support for the muscle fibres and whole muscle. They also hold nerves, blood and lymphatic vessels. These sheaths are:

- **The epimysium** ("epi-" = upon, "myo-" = muscle) encloses the entire muscle.
- **The perimysium** ("peri-" = around) extends from the epimysium into the muscle, and encloses bundles of muscle cells. Each bundle of muscle fibres is called a fascicle.
- **The endomysium** ("endo " = inner) covers individual muscle fibres.

A single muscle cell (myofibre) contains many myofibrils – these are cylindrical bundles of contractile proteins that lie parallel to each other. The length of skeletal muscle cells means that they are often referred to as muscle fibres. Each myofibril contains two contractile proteins (known as microfilaments), which overlap each other.

The thick microfilaments contain the protein myosin, and the thin microfilaments contain the protein actin. The interlocking microfilaments of actinomycin give the myofibrils their striped (or striated) appearance.

Contract and Relax

The contraction of a skeletal muscle usually causes a bone to move, but normally only the bone to which one end of the muscle is attached will move. This is called the insertion end of the muscle; the stationary end is called the origin.

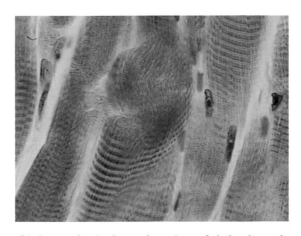

This image clearly shows the stripes of skeletal muscle.

Skeletal Muscle

Your muscle cells use glucose or fatty acids as their fuel, and this releases both heat and energy in the form of ATP. ATP is used to power the movement of actin and myosin microfilaments, which slide past each other to shorten the cell, and so cause contraction. The whole process is started by the nervous system. A signal from a motor nerve causes the release of calcium, which floods into the muscle cells and triggers the contraction of myofibril. This shortens the muscle, and moves the part of the body to which the muscle is attached.

To relax, the actinomycin microfilaments slide apart again, and the muscle becomes longer.

Muscles vary greatly in size according to how many muscle fibres are present, the diameter of the individual fibres, and the length of the muscle belly. These are all features related to the role of the particular muscle. For a given muscle, its size is, to a large extent, predetermined because of the length of its belly, the number of muscle fibres and the amount of work that muscle is persistently required to do.

Working Out

The diameter of individual muscle fibres, and therefore the muscle itself, may be altered by the work performed by the muscle. For example, if you regularly lift heavy weights, this stimulates protein production in the working muscles, such as the biceps. Over time, this leads to a bulking-up of its muscle mass. On the other hand, if you do not use the biceps muscle regularly then muscle mass decreases. So it really is a case of "use it or lose it". Similar processes also operate on other types of muscle. For example, the heart muscle enlarges if you undertake regular exercise so it can provide a more powerful contraction, forcing more blood out at each contraction (greater stroke volume). This is why athletes have a lower heart rate than those who perform little or no exercise.

Understanding the relationship between workload and muscle bulk helps us to partly understand why postural muscles in the body are so large. When you're standing, these muscles must support the joints that are particularly subjected to gravity. The bulk of these muscles provides a reminder of how significant this force is. For example, the erector spinae muscles of the back need to be large in order to keep the back stable and maintain an erect vertebral column and head. The gluteus maximus of the buttock has to be bulky to stabilise the hip joint against gravity, and so facilitate an erect posture. And finally, the quadriceps femoris of the thigh has to be substantial to keep the knee joint stable against gravity so that the leg can extend or straighten.

The Skin, Skeleton and Muscles

Muscles at Work

When muscles contract, they can only pull, not push. Muscles, therefore, have to work in pairs. The muscle that produces movement is called an agonist, and the muscle that opposes the movement is called an antagonist.

The antagonist does not prevent total movement, and only opposes contraction sufficiently to protect the joint involved. The agonist muscle may extend (straighten) or flex (bend) a joint, which is why the muscles that perform these actions are known as either extensors or flexors.

Let us consider the movement of the elbow joint during bending and straightening the arm. The muscles involved are the biceps and the triceps. The triceps is an extensor muscle as its contraction extends the arm. The biceps are flexors, and cause the arm to flex, or bend, at the elbow joint when they contract. During the straightening of the arm, the joint extensor muscle (triceps) will be agonistic as the arm is extended, and the flexor biceps is

antagonistic. When the elbow is flexed, the situation is reversed.

A "Pulled Muscle"

A pulled muscle, or muscle strain, occurs when excessive force, overuse or stretching causes tearing of muscle fibres. When skeletal muscle fibres have formed, their cells cannot divide, so the healing of a "torn" muscle will create scar tissue. Being less elastic than normal muscle tissue, the scarred muscle will lose some of its efficiency and, if the damage is substantial and immobility prolonged, scar tissue may even become ossified ("turned to bone"), reducing elasticity even more. This is called myositis ossificans.

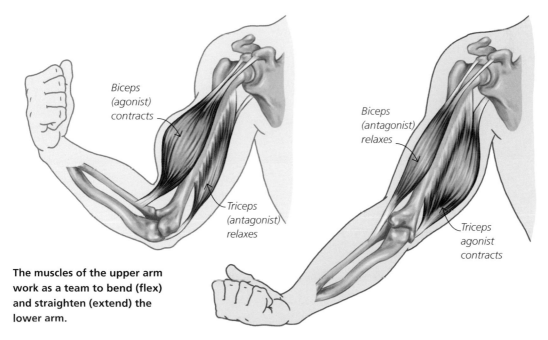

Biceps
(agonist)
contracts

Triceps
(antagonist)
relaxes

Biceps
(antagonist)
relaxes

Triceps
agonist
contracts

The muscles of the upper arm work as a team to bend (flex) and straighten (extend) the lower arm.

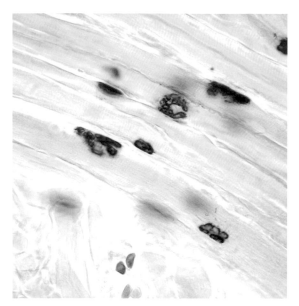

When a signal arrives at the neuromusclar junctions (the dark spots in this micrograph), it stimulates the release of acetylcholine and causes excitation (depolarization) of the membrane surrounding the muscle cell, known as the sarcolemma. The electrical current generated within the sarcolemma is then transmitted and starts the release of calcium ions from intracellular stores, and this calcium triggers the muscle contraction.

Stimulation of Muscle Fibres

The nerve cells that cause muscles to contract are known collectively as motor neurons. The motor nerve activity will frequently have been generated within the brain to pass to muscles of the face and skull from the brain (via cranial nerves), or to the muscles of the body via the spinal cord (via spinal nerves). These motor neurons terminate at the muscle fibre in a structure referred to as the neuromuscular junction.

The neuromuscular junction or gap between the motor neuron and muscle cell is a synapse. A muscle fibre forms numerous synapses along its length with terminals from the same nerve cell. So, if the nerve cell is active, impulses directed along the terminals will stimulate the entire fibre length at the same time. This makes sure that the whole fibre contracts. The synapse at the neuron-muscle fibre interface is somewhat different to that of a neuron-neuron synapse, and it is called a motor end plate. However, the end plate functions in much the same way as other synapses as it is a chemical neurotransmitter.

Muscle Cramp

The strength of a muscle contraction will depend on the rate or level of nerve excitability or stimulation and the local chemical environment. This is influenced by the length of time that a contracting muscle has had to spend in anaerobic conditions, when oxygen availability is low. Spasm and cramp occur when the excitability is increased to the point that nerve stimulation of the muscle creates a sustained powerful contraction, leading to low oxygen levels (hypoxia) and consequently, pain. Cramp is normally associated with a prolonged period of physical activity when the muscle is receiving a high frequency of nerve stimulation that leads to tetany – a sustained muscle contraction. Spasm is usually associated with lower-level muscle activity and can occur, for example, simply when getting up from a sitting position! The spasm and cramp are protective and they prevent muscle damage occurring through sustained use – but this doesn't make them any less painful!

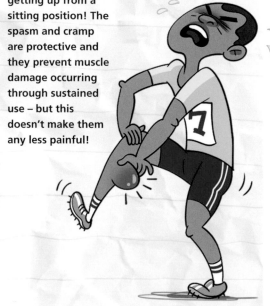

The Skin, Skeleton and Muscles

Chapter 3
THE CARDIOVASCULAR SYSTEM

Delivery System

The cardiovascular system consists of the heart, the system's pump, and the blood vessels, the delivery routes. The system is designed to keep blood flowing, transport nutrients, oxygen and hormones to the cells, and remove waste, such as carbon dioxide.

The cardiovascular system does not work on its own. It needs the co-operative activities of other systems to maintain blood composition and preserve homeostasis in the body. The digestive, respiratory, lymphatic and urinary systems, for example, maintain the homeostasis of blood. The involuntary (autonomic) nervous system and endocrine systems co-ordinate cardiovascular (and other system) activities. Their reliance on each other to preserve homeostasis means a disturbance in one results in a malfunction of another.

Naming Blood Vessels

The names of blood vessels, just like names of bones as shown in chapter 2, are complex, but the name of a blood vessel usually gives a clue to its appearance and/or distribution to a particular organ. So, if you become familiar with the major skeletomuscular and neural "landmarks" of the human body, there should be few surprises.

At the heart of the system is the heart, if you pardon the pun! This organ never stops pumping blood at regular pulses through the network of vessels to collect oxygen and nutrients that are delivered to the body's cells, and waste products that are carried away from body's cells for their removal from the body.

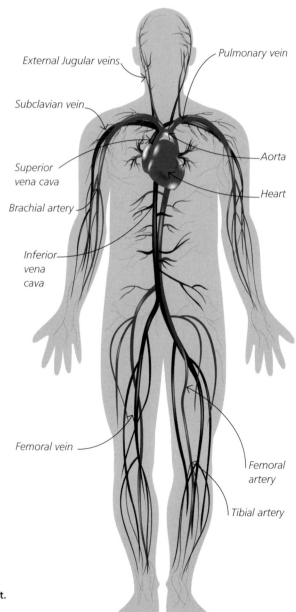

External Jugular veins

Pulmonary vein

Subclavian vein

Superior vena cava

Aorta

Heart

Brachial artery

Inferior vena cava

Femoral vein

Femoral artery

Tibial artery

The cardiovascular system, including the heart, the arteries and veins of the human body, form a complete network that carries blood to nearly every cell in the human body. Every day, about 7,500 litres (13,200 pints) of blood is pushed around the blood vessels by the heart.

Blood Vessels

The cardiovascular system is a "closed" system of elaborate tubings and the heart pumps blood throughout this system in one direction.

The structure of the blood vessels varies according to what the vessel does, but all of them, except capillaries, have the same basic structure of three layers, known as coats, or tunicae:

- **The innermost tunica interna** consists of a single endothelial layer of flattened cells. Capillaries are composed only of this layer, so as to aid their rapid exchange of materials between the blood, interstitial fluid and cells of the body.
- **The tunica media** is the middle layer and is made up mainly of elastic and smooth (involuntary) muscle fibres.
- **The tunica externa** is the outer connective tissue sheath.

There are five types of blood vessel in the cardiovascular system that work as a secret team and are responsible for transporting blood and its components to and from the cells of the body. These vessels are arteries, arterioles, capillaries, venules and veins.

The Arterial System

Arteries always transport blood away from the heart. They usually carry oxygenated blood, except for the pulmonary arteries and the umbilical artery in the developing foetus. These circulations carry deoxygenated blood (that is, blood with a low oxygen content). Arteries are either elastic or muscular arteries. Elastic arteries, such as the aorta, pulmonary arteries and their major branches, have larger diameters. Their middle layer contains more elastic fibres. This allows for "stretching", to accommodate the blood volume it receives from

the heart when its lower chambers (ventricles) contract, and recoiling, when the heart muscle relaxes. Elastic arteries are sometimes referred to as the "conduction vessels", since they conduct blood away from the heart to the muscular arteries. These arteries are medium-sized vessels whose middle layer contains more smooth muscle fibres. They can adjust blood flow by contraction or relaxation of their muscle fibres to suit the needs of the cells supplied. Sometimes they are referred to as the "distribution vessels" as they distribute blood to

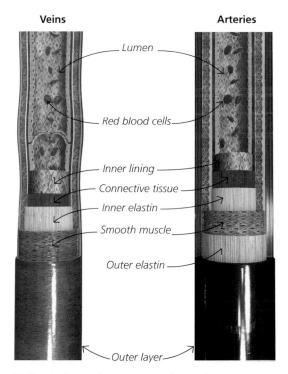

Both arteries and veins are made up of various layers of connective tissues and muscle.

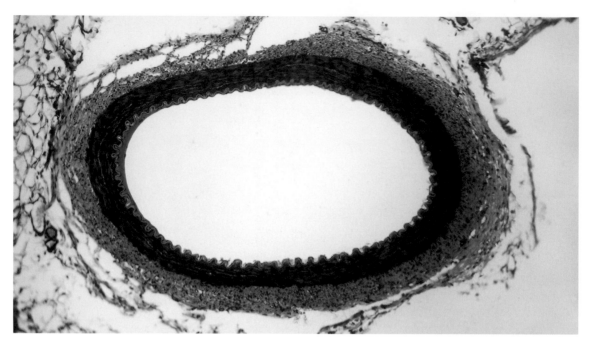

This slice through an artery clearly shows the thick muscular walls (stained darker). The larger arteries can have a diameter of 10 mm (²/₅ in) or more.

the arterioles, which are the smallest arteries in the arterial system.

Arterioles deliver blood to the capillary networks, so that these networks can supply the cells with the blood components that they require. The arterioles closest to the capillaries contain precapillary sphincters (circular muscle fibres). These muscle fibres can alter the diameter of the arterioles – partial contraction reduces blood flow and complete contraction stops blood flow. As such, these sphincters regulate blood flow to the capillaries so the tissue cells they supply depend upon their needs. For this reason, the arterioles are called the precapillary resistance vessels.

Capillaries

A single arteriole divides into several capillaries, which in turn collect to form many venules. Typically, a capillary consists of a tube made from a single layer of cells (equivalent to the tunica interna). Capillaries are usually the width of a single blood cell and this reduces the blood flow rate to

600 times slower than it is in the aorta. They are found close to almost all body cells, hence their main activity is to permit the exchange of materials between blood and cells, which is why they are sometimes called the "exchange vessels". The number of capillaries depends upon the activity of the tissues they supply. For example, high activity muscle and nervous tissues have a rich capillary supply, whereas low activity tissues, such as tendons, have a poor capillary supply.

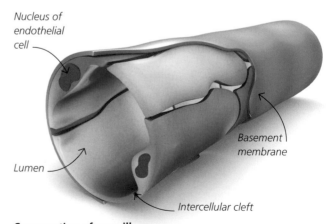

Nucleus of endothelial cell

Basement membrane

Lumen

Intercellular cleft

Cross-section of a capillary

The Cardiovascular System

The Venous System

The venous system is sometimes referred to as the body's "drainage system" – it drains blood from the capillaries towards the heart. On their way back from the capillaries, the venous vessels increase in diameter, their walls thicken, and they progress from the smallest veins (venules) to larger veins (such as the gastric vein), and finally to the largest veins (the superior and inferior vena cavae).

The walls of veins are thinner than arterial walls, since they have less elastic tissue and smooth muscle. Their larger lumens mean they offer less resistance to blood flow. This is essential as the blood pressure within the venous circulation is low. To help blood return to the heart, there are valves in the deep veins of the limbs, especially the legs, where blood must travel a considerable distance against gravity when a person is standing upright.

Valves

When viewed under a microscope, valves appear as small flaps that project inwards from the vein walls. When blood is being pushed through the veins, the valves are pushed open. Then, when blood pressure drops, blood attempts to move backwards, closing the flaps of the valves. The flaps are also controlled by the "squeezing" (contractile)

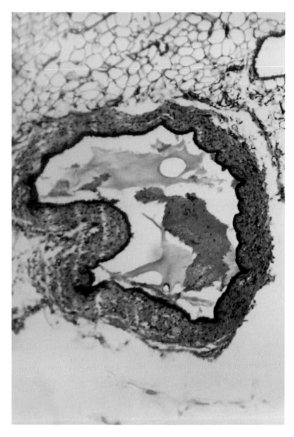

Compared to the artery on the left, this vein has thinner walls with less muscle tissue. The largest veins can measure more than 13 mm ($^1/_5$ in) across.

Secrets About Blood Vessels

- **The circulatory system is extremely long, its length is approximately 100,000 km (62,000 miles) enough to wrap around the Earth about 2.5 times.**

- **Blood vessels are affected by the weather. Vessels that are close to the skin's surface expand to release heat, allowing you to cool down, and constrict when cold to save heat.**

- **Veins are not blue as they appear in many diagrams, instead they are dark red because the blood they hold has very little oxygen.**

- **Blood in the arteries is bright red because it holds lots of oxygen.**

- **Blood vessels change with age and time, and damage can start early, even during childhood. Obese teens with high blood pressure may show signs of thicker arteries by the age of 30.**

- **Smoking causes immediate damage to the lungs, and also progressively to the blood vessels. Chemicals in smoke can be a major risk factor in people who have had a heart attack or a stroke.**

action of the skeletal muscles that surround the veins. These valve movements ensure that blood flows towards the heart and not backwards. Valves are absent in small veins and very large veins in the chest and abdominal cavities. Veins contain about 60 per cent of the blood volume, which is why they are also known as the "blood reservoirs".

Varicose Veins

Valve damage may be congenital (present at birth) or it can develop following birth and may occur if the venous system is exposed to high pressure for long periods, for example in pregnancy, obesity, and when you stand for long, extended periods. As a result, damaged valves become "leaky" (or incompetent) and they allow a backflow of blood, causing blood to pool below the affected valve.

These veins that have damaged valves are called varicose veins and they can become long and tortuous. Consequently, fluid may leak into the surrounding tissues, producing a swollen appearance (oedema). The affected vein and tissue around it may become inflamed and painfully tender, and varicose ulcers may develop. Areas next to varicose veins may ache, because the diffusion of oxygen and nutrients following oedema is reduced. These veins are liable to bleeding (haemorrhage) if knocked, but the patient's main complaint is usually of their appearance. Veins close to the surface of the skin are more likely to develop into varicose veins. Deeper veins are not usually as vulnerable, because surrounding skeletal muscle prevents their walls from stretching too much.

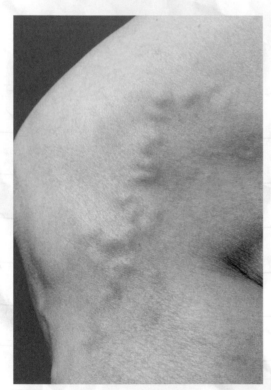

Varicose veins appear as twisted blue blood vessels close to the surface.

While normal vein valves prevent the backflow of blood, leaky, incompetent valves allow some blood to flow back, causing the vein to swell.

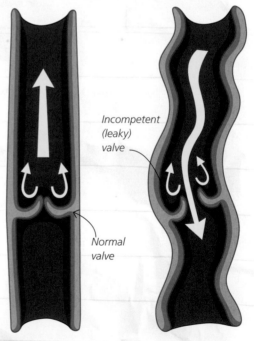

Incompetent (leaky) valve

Normal valve

Normal vein

Varicose vein

The Cardiovascular System

Dual Circulation

There are two main parts to the circulatory system – the systemic circulation and the pulmonary circulation.

The systemic circulation pumps oxygenated blood from the left side of the heart out through the systemic arterial blood vessels to the tissue cells. Here oxygen and nutrients are passed into cells and carbon dioxide and other wastes are passed from cells into blood. Systemic veins carry this deoxygenated blood (blood low in oxygen content) from the cells back to the right side of the heart, from where it is pumped to the lungs via the pulmonary circulation.

The pulmonary circulation is where the deoxygenated blood that has circulated in the venous system is reoxygenated.

Main Veins

These are usually coloured blue in diagrams, and take blood back to the heart. However, pulmonary veins are coloured red because they return oxygenated blood from the lungs back to the left side of the heart.

Femoral vein – contains special valves to help maintain blood flow back to the heart against gravity.

Inferior vena cava – is the largest vein in the body and carries deoxygenated blood back to the right side of the heart from blood vessels, tissues and organs below the heart.

Superior vena cava – carries deoxygenated blood back to the right side of the heart from blood vessels, tissues and organs above the heart.

Jugular veins – drain blood from the head and brain.

Subclavian vein – collects deoxygenated blood from the arm.

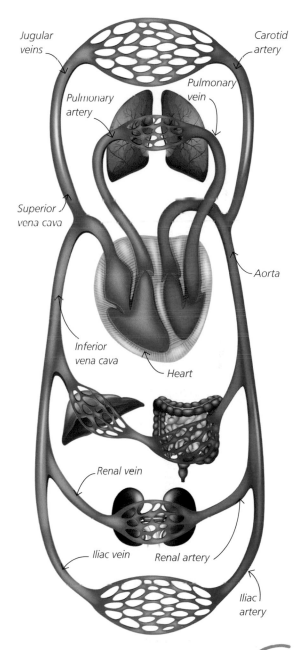

Jugular veins

Carotid artery

Pulmonary vein

Pulmonary artery

Superior vena cava

Aorta

Inferior vena cava

Heart

Renal vein

Iliac vein

Renal artery

Iliac artery

Pulmonary arteries transport deoxygenated blood from the right side of the heart to the lungs. Here, the carbon dioxide is passed into the airways from blood and the blood is replenished with oxygen from the lungs. Pulmonary veins now transport this oxygenated blood back to the heart. The pulmonary circulation is the only part of the adult circulation in which arteries carry deoxygenated blood and veins transport oxygenated blood.

Coronary Circulation

The heart chambers are continuously bathed with blood, which provides nourishment to the innermost cells of the heart. The middle layer's muscle cells and outer cells, however, are too far away to receive nutrients from this blood source. These cells need their oxygen and nutrients from their own blood vessels – the coronary circulation. Strictly speaking, this circulation is part of the systemic circulation, since it transports oxygenated blood from the left side of the heart and returns deoxygenated blood to the right side of the heart.

It consists of two main arteries, the right and left coronary arteries, which are the first branches of the main artery of the body, the aorta, which leaves the left side of the heart. These arteries are subdivided into a complex network of smaller and smaller blood vessels that supply the heart

The dual circulation system sees blood flow to the lungs, then to the body and then back to the lungs.

muscle with its oxygen and nutrients. As in the rest of the body, blood that has given up its oxygen is transported back to the heart through the coronary veins, which empty it into the right side of the heart through the coronary sinus.

Coronary Arterial Disease

Coronary arterial (heart) disease (CAD) is caused by a lack of oxygen to the heart's muscle, arising from a narrowed or blocked coronary artery. Atherosclerosis is the most common cause of CAD. This is a type of "hardening of the arteries" where the artery becomes blocked. The blockage, known as a plaque, is largely made up of cholesterol. Factors that can lead to atherosclerosis include family history of CAD, lack of exercise, diabetes, stress and smoking. A blood clot (thrombus) can also block one of the coronary arteries.

Some people who have had a small heart attack may have had what is referred to as a "silent" myocardial infarction and they are unaware that they have had a coronary. Others may experience chest pain (angina), while people who have a large heart attack may die because the heart goes into severe failure.

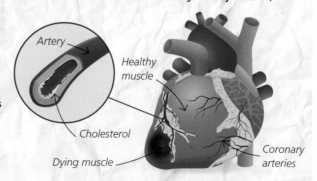

Artery

Healthy muscle

Cholesterol

Dying muscle

Coronary arteries

The Heart

Your heart is a powerful muscular pump, which is located between the lungs, slightly tipping to the left in the chest cavity (thorax).

Situated in front of the spine (vertebral column) and behind the chest bone (sternum), your heart is often compared with the person's closed fist to demonstrate its approximate, but variable size.

The heart contains three walls. The outermost wall is a supportive and protective double fluid-containing wall called the pericardium. The middle wall of the heart is a specialized form of muscle called cardiac muscle (myocardium). The innermost wall of the heart is the endocardium and this has a smooth lining to stop blood from clotting.

Two Halves

Your heart is divided into two halves; right and left. The right receives deoxygenated blood from the systemic circulation and the left, oxygenated blood from the pulmonary circulation. Each half is further divided into two communicating chambers – an upper atrium (plural, atria) and a lower ventricle, through which blood is pumped. The atria receive blood entering the heart from blood vessels. The ventricles push the blood out, and form the actual "pumps" of the heart. The atria only pump blood into the ventricles, so their walls are relatively thin. The right ventricle pumps blood to the lungs and the left ventricle pumps blood around the body against considerable pressure, so it has a thicker, more muscular wall than the right ventricle. The chambers of the heart are separated by a dividing wall of tissue called the septa.

The atria, ventricles and vessels leaving the heart are separated by valves, which function in the same way as the valves in veins – ensuring one-way flow, and preventing unwanted backflow. The right ventricle is separated from the right atrium by the tricuspid valve. The left atrium is separated

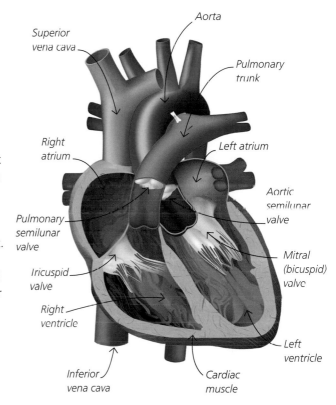

Each half of the heart has an upper atrium and lower ventricle, separated by valves. Each of these heart chambers holds about 70 ml (2 fl oz) of blood.

from the left ventricle by the mitral or bicuspid valve. The pulmonary valve houses the entrance to the pulmonary trunk, which takes deoxygenated blood from the heart to the lungs. The aortic valve houses the entrance to the aorta, which takes oxygenated blood from the heart to the rest of the body.

Secrets of the Heart

- The study of the human heart and its disorders is known as cardiology.
- The heart beats on average 70 times a minute; that equates to 100,000 beats a day; 3.6 million beats a year; and 2.5 billion beats in your lifetime.
- Each heart beat takes about 0.8 seconds. The atrial systole accounts for 0.1 seconds and ventricular systole for 0.3 seconds; the remaining 0.4 seconds is the period of total heart relaxation (diastole).
- The heart can pump volume 5 litres (9 pints) of blood per minute at rest to 30 litres (53 pints) per minute during exercise.

- The pressure within the left ventricle could squirt blood about 10 meters (33 ft) into the air.
- The heart rate is regulated by sympathetic and parasympathetic nerve stimulation, which speed up or slow the heart down, respectively, as and when required.
- The sound of the heart valves closing produces the familial sounds "lub-dup" when you listen to the heart with the stethoscope.
- Your coronary circulation accounts for 1/12th of the total output from the heart, even though the heart represents only 1/200th of the body's weight.

Starting Signal

Your heart's muscle, like any muscle requires a nervous impulse to enable it to contract. This impulse is conducted through the heart's muscle via special nerve fibres. A natural "pacemaker" called the sinoatrial node, generates regular electrical impulses that cause the heart muscle to contract in a controlled sequence of events, known as the cardiac cycle, that occur with each heartbeat. There are three stages or phases to each heartbeat: resting, atrial contraction, then ventricular contraction stages.

During the resting phase (diastole), the right side of the heart fills with the deoxygenated blood (from the body) and the left side of the heart fills with oxygenated blood (from the lungs). About 70 per cent of the blood the atria receives passively trickles into the ventricles through the opened atrioventricular (tricuspid and mitral) valves. During the second stage (atrial systole), the two atria contract at the same time to squeeze the remaining blood into the two ventricles. During the third phase (ventricular systole), both ventricles contract at the same time and the changes in pressure causes the closure of the atrioventricular valves and the opening of the semilunar (pulmonary and aortic) valves pumping the blood into the lungs

from the right ventricle and through the body from the left ventricle. When the ventricles are empty, the heart muscle relaxes (diastole) and the cycle begins again.

The electrical impulse flowing through the heart triggering the heartbeat can be recorded to produce a tracing called the electrocardiogram (ECG), which can be used to identify normal or abnormal heart rhythms.

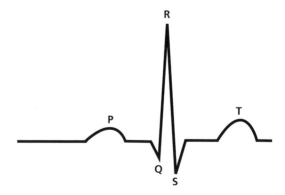

The readout from a normal ECG produces various waves (PQRST). Each wave represents the electrical activity which is needed to produce the heartbeat.

The Cardiovascular System

Blood Pressure

Blood circulates because the heart can push the blood at great pressure. Blood pressure is the measure of how strongly your blood presses against the walls of your blood vessels.

The highest pressure, created by the left ventricular pump, is recorded in the aorta before its coronary branches. The lowest pressure is at the junction of the superior and inferior vena cavae. The average pressure is most important, since the left ventricle pumps blood in a pulsating manner and tissue flow generally varies accordingly.

Systemic Pressure

The normal systemic arterial pressure in a resting young adult is about 120 mmHg and 80 mmHg (15.99 kPa and 10.66 kPa). The higher value is observed following the removal of blood from the left ventricle during systole, and is therefore called the systolic pressure. The lower value is observed at the end of diastole phase, and is therefore called the diastolic pressure.

You should remember that there is no such value as a "normal blood pressure" reading for the population as a whole. However, there is a normal value for any particular individual, but even that value will vary according to the different activities they undertake, and the age of the person. For example, at rest your blood

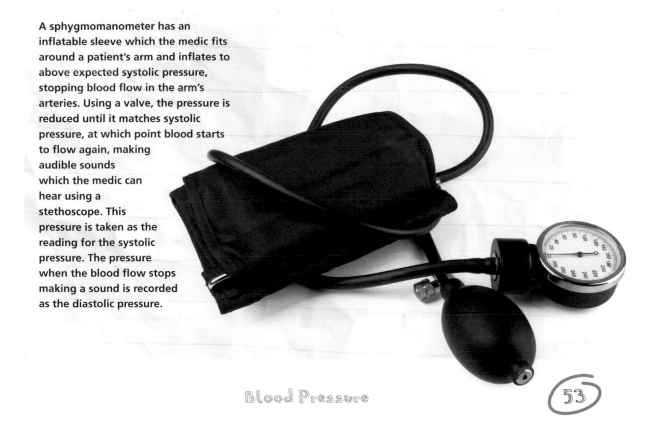

A sphygmomanometer has an inflatable sleeve which the medic fits around a patient's arm and inflates to above expected systolic pressure, stopping blood flow in the arm's arteries. Using a valve, the pressure is reduced until it matches systolic pressure, at which point blood starts to flow again, making audible sounds which the medic can hear using a stethoscope. This pressure is taken as the reading for the systolic pressure. The pressure when the blood flow stops making a sound is recorded as the diastolic pressure.

pressure is lower than when you are taking part in strenuous exercise. Also, when you advance in years, the systolic blood pressure rises; averaging about 90 mmHg (11.99 kPa) at one year of age to about 105 mmHg (13.99 kPa) at 10 years.

Readings that are higher than these are referred to as high blood pressure, and sustained high levels can have adverse effects on the body (see below).

In the USA, it is believed that about 75 million people (29 per cent of adults) have high blood pressure and, according to the Centers for Disease Control and Prevention, conditions caused by this cost the nation $46 billion each year.

High Blood Pressure and Obesity

High blood pressure (hypertension) does not have any symptoms, so the only way to know if you have it is to have your blood measured regularly by your doctor or nurse. Many doctors recognize hypertension if there is a persistent resting systolic pressure which is greater than 140 mmHg or 18.66 kPa (in elderly people greater than 160 mmHg, or 21.33 kPa) and diastolic pressure of over 90 mmHg (11.99 kPa). Clinicians are concerned with blood pressure readings because a significantly increased mortality exists in people who have hypertension. The risk increases rapidly with increasing pressure, and the patient is most at risk from a stroke (cerebral vascular accident, CVA) or a heart attack (myocardial infarction, MI).

We do not know exactly what causes hypertension, but we do know that your lifestyle can affect your risk of developing it. You are at risk if you eat too much salt, you are not active enough, you drink too much alcohol, you don't eat enough fruit and veg, and if you are overweight. High blood pressure and obesity frequently coexist. Current evidence suggests that it is fat within the abdomen ("your beer belly") that has the most significant impact on your blood pressure. If your waist size is equal to or more than 39 cm (35 in) in women and equal to or more than 102 cm (40 in) in men, it increases your risk of hypertension, and also cardiovascular disease, and diabetes.

Pulse

The pulse can be felt where an artery lies near the skin's surface and a bony or firm surface. It represents the expansion and elastic recoil of an artery with each left ventricular contraction (systole).

So you can see why the strongest pulse is in the arteries, closest to the heart, and that it weakens as it passes through the arterial system, disappearing altogether within the capillary networks. The most common and accurate pulse taken is the wrist or radial pulse. However, other areas, such as the carotid (neck) pulse, can be used instead if the radial pulse is difficult to find, which sometimes occurs if the arm is obese.

Racing Pulse

Three factors are observed when recording a person's pulse: rate, strength and rhythm. The pulse rate equates to the heart rate and is calculated by counting the beats felt within 60 seconds. Pulse rates vary with age, level of activity, level of fitness, body temperature, if you have an infection or illness and even when you are scared or anxious. The resting pulse rate should be between 60 and 80 beats per minute (bpm), when an adult is lying quietly in bed. An abnormally fast pulse of 100 or more bpm is referred to as a tachycardia and is considered normal when exercising, but can also be a sign of infections, fever, emotional upset, pregnancy, anaemia, cardiac failure or an overactive thyroid. Whereas an abnormally slow pulse of below 50 bpm is called a bradycardia and occurs when the impulse from the heart's pacemaker (SA node) does not always reach the ventricles (as in heart block) and the ventricular contraction rate slows down. It may indicate that you have an underactive thyroid, a low body temperature (hypothermia), or that the pacemaker has a low oxygen content. It can also occur in response to drug therapy as occurs in some people

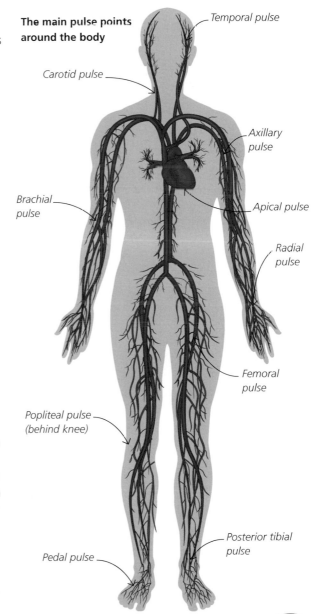

The main pulse points around the body

Temporal pulse

Carotid pulse

Axillary pulse

Brachial pulse

Apical pulse

Radial pulse

Femoral pulse

Popliteal pulse (behind knee)

Posterior tibial pulse

Pedal pulse

The radial pulse can be detected by pushing two or three fingers lightly against the centre of the underside of the wrist.

who are prescribed beta blockers. Extreme bradycardia and tachycardia result in inadequate filling of coronary arteries, which can lead to a heart attack and/or a lack of oxygen to the brain, which can cause confusion and disorientation, and can give rise to brain damage.

Strong or Weak?

The strength of the pulse is important because it can give an indication of the heart's output and probable blood pressure. The strength of the pulse is reliant upon the force of left ventricle contraction and the amount of blood pumped out on this contraction. There is a clear relationship between the pulse felt at different sites and systolic blood pressures. For example, the radial pulse is greater than 80 mmHg and the carotid is greater than 60 mmHg. A weak pulse indicates that a poor volume of blood is being pumped out with each ventricular contraction, or that a blood vessel may be blocked through arteriosclerosis above the pulse site.

The rhythm of the pulse is the pattern in which the beats occur. In healthy people, the pattern is regular because the chambers of the heart are contracting in a coordinated manner, while an irregular pulse may suggest an underlying disorder.

Pulse Irregularity

Most people reading this book will have experienced palpitations (heart beats felt by the patient on the chest wall) in their lifetime and these are often "normal", and may occur during extreme exercise, severe emotion or they may also be associated with an irregular heart rhythm (arrhythmias). Incidences of palpitations should be reported to your doctor, but the doctor might not treat it, if it is believed to be an intermittent irregularity of rhythm. However, if the doctor believes it is a regularly occurring irregularity, then it could be the result of a heart block. Otherwise an irregular pulse rate may indicate atrial fibrillation, the most common irregular cardiac rhythm. It occurs in 2 to 4 per cent of the adult population over the age of 60 years.

Chapter 4
INTERNAL PROTECTORS: WARRIORS BEHIND THE SCENES

Red Blood Cells

Blood acts as a link between the outside and the body's tissue fluid and cells. Its importance cannot be underestimated as any body part deprived of blood can't function properly and will die in a matter of minutes if the circulation is not restored!

When you cut yourself, blood appears to be a uniform, dark red liquid that, if left to stand for a few minutes, normally clots (solidifies). However, if you were to look under a microscope you would see that blood is a mixture of cells, floating in a pale yellow fluid called plasma!

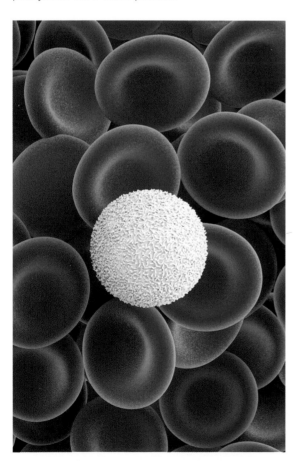

Red blood cells, or erythrocytes, swarm around a white blood cell. Red blood cells usually last for about 120 days before being broken down by the body.

Plasma is largely made up of water. The remaining constituents of plasma include: clotting proteins, enzymes, hormones, nutrients (such as glucose, amino acids, fatty acids) cholesterol, waste products (for example urea), and dissolved salts (electrolytes), such as calcium. These components make this fluid sticky, and resistant to flow. There are three main types of blood cell:

- **Red blood cells** (erythrocytes)
- **White blood cells** (leucocytes)
- **Platelets** (thrombocytes)

Oxygen Carriers

The erythrocytes are much more numerous than the other blood cells. Their biconcave, or "doughnut", shape provides a large surface area so it can quickly exchange the respiratory gases – oxygen for carbon dioxide at the cells, and carbon dioxide for oxygen in the lungs. The cytoplasm of red cells contains haemoglobin, a red pigment whose main function is the transportation of the respiratory gases, primarily oxygen.

Erythrocytes also define a person's blood group, since their membrane contains chemical markers, called antigens. Worldwide, there are more than 35 blood groups. Although these groups are of immense importance in forensic medicine, such as identifying criminals and corpses from different regions of the world, only two principal blood group systems, the ABO and Rhesus systems, are important clinically. This is because transfusion of an inappropriate type of blood can promote clumping (agglutination) of red cells in the recipient and can result in symptoms such as pains in the chest, back and abdomen, difficulty in breathing, jaundice, blood in the urine, chills and fevers.

White Blood Cells

Leucocytes (white blood cells) are a part of the body's immune system. As they circulate in blood and lymph vessels and accumulate in lymph nodes they are "looking" for signs of pathogenic invasion in these areas and adjacent tissues.

Leucocytes have a nucleus and do not have a coloured pigment, so they appear "white" in contrast to red blood cells. Leucocytes fall into two main groups: granulocytes and non-granulocytes.

Granulocytes

The cytoplasm of granulocytes contains granules in their cytoplasm, while non-granulocytes do not have these granules. Granulocytes, called neutrophils, eosinophils and basophils, are classified according to the reactions of their granules to staining techniques used in hospital haematology departments.

Neutrophils are so named since neutral dyes stain their granules purple. Their granules contain enzymes and other chemicals that kill bacteria. They are very mobile and are the first white cells to arrive at a site of any damage or injury. They are also the most active phagocytes in response to tissue destruction by bacteria, so it is inevitable

that large numbers of neutrophils are often killed themselves in any bacterial infection. These dead cells (white cells and dead bacteria) make up the pus that occurs at an injured site.

Eosinophils are stained by acidic dyes and are phagocytic to parasites and bacteria. These cells also combat irritants that cause allergies, which is why their numbers increase in response to allergic reactions and parasitic infections. Their granules contain powerful enzymes and, because they are involved with antibody immune responses, they function to neutralize the effects of inflammatory substances, such as histamine, released by damaged tissue. Eosinophils especially accumulate at the site of allergic reactions, for example in the nose of a person experiencing a bout of hay fever.

Basophils are so called because their granules stain easily with basic dyes. These cells are important in allergic reactions. When they enter inflamed tissues, basophils bind the irritant

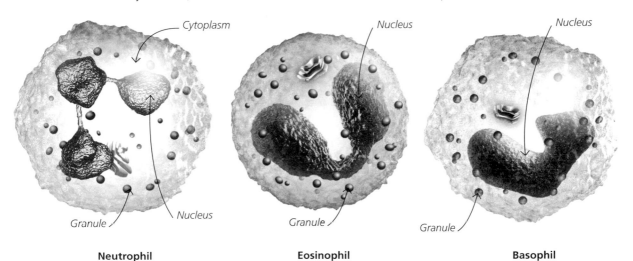

Neutrophil   Eosinophil   Basophil

chemical to their cell membrane. The binding encourages basophils to release the contents of their granules – histamine and serotonin. Both of these dilate the capillary blood vessels and increase their permeability, so blood components can pass through to the inflamed area. The movement of these blood components may be observed as a localized swelling called oedema. Itching and pain are also associated with large secretions of histamine into the inflamed tissue. Other chemicals released by activated basophils, sometimes called mast cells, attract phagocytic eosinophils and even more basophils to the affected area. Their job is to engulf bacteria and other microbes and their particles.

All granulocytes have a lobed nucleus, whereas non-granulocytes contain either a kidney-shaped nucleus (these white cells are called monocytes) or a sphere-shaped nucleus (these white cells are called lymphocytes).

Non-Granulocytes

Monocytes, also called macrophages are large phagocytic white cells. They exist as either:

• **Free monocytes** (or "mobile" macrophages) located outside the blood,
Or
• **Immobile (or fixed) monocytes** found only in connective tissues.

Migratory monocytes, or mobile macrophages, are very mobile (hence their name), so they have plenty of mitochondria in order to produce the massive amounts of energy needed for their enhanced mobility. Macrophages arrive at an injury quickly to establish their phagocytic function. These cells also release chemical attractants to bring even more macrophages and phagocytes to the inflamed area. Monocytes entering the inflamed area are referred to as "scavenger" macrophages, because they clean up any debris following an injury, in much the same way as vultures pick at the remains of another animal's kill! The migratory monocytes move into the bone marrow, spleen, liver and lymph nodes where they develop into larger specialized cells, such as the liver's Kupffer cells, which play a key role in the destruction of red blood cells that have come to the end of their life.

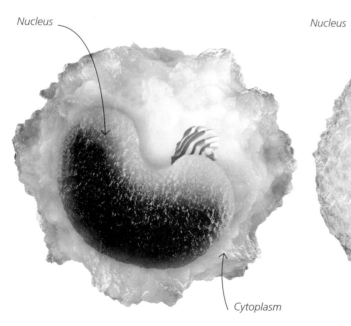

Nucleus

Cytoplasm

Monocyte

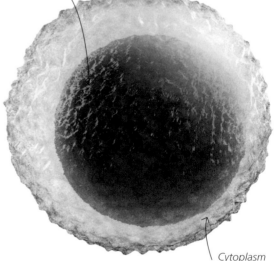

Nucleus

Cytoplasm

Lymphocyte

Internal Protectors: Warriors Behind the Scenes

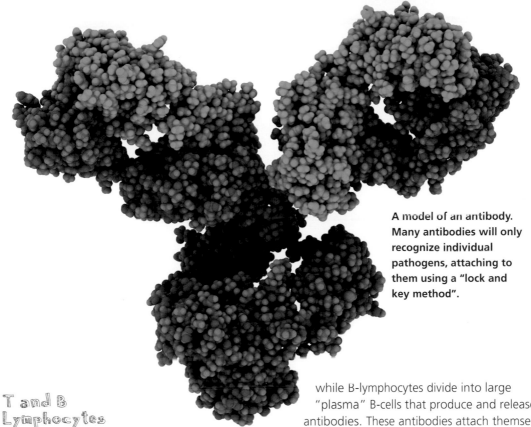

A model of an antibody. Many antibodies will only recognize individual pathogens, attaching to them using a "lock and key method".

T and B Lymphocytes

Most lymphocytes are housed inside the lymphatic system, hence their name, although some appear in blood, usually when there is an infection. Lymphocytes are categorized as either T-lymphocytes (or T cells) and B-cells. Lymphocytes are also divided into a number of subclasses. T-lymphocytes are divided into T-cytotoxic (killer) cells, which kill microbes and tend to survive to go on to kill all the other microbial targets, T-helper cells, which stimulate the B-lymphocytes to produce antibodies, T-delayed hypersensitivity cells, which are involved in some hypersensitive or allergic reactions, and T-suppressor cells, which bring antibody production to a halt when the individual has overcome the infection. B-lymphocytes are divided into either large plasma antibody-producing cells or B-memory cells.

In summary, lymphocytes have several different roles. The T-lymphocytes attack any potential pathogens directly in the cellular immune response, while B-lymphocytes divide into large "plasma" B-cells that produce and release their antibodies. These antibodies attach themselves to foreign antigens in a "lock and key mechanism" rendering the antigen harmless to body tissues, and therefore helping combat infection and giving the body immunity to some diseases. There are several types of antibodies, or immunoglobulins ("Ig"s):

- **IgA** is mainly found in secretions that protect internal surfaces, and are found in tears and the respiratory and intestinal tracts.
- **IgD** is mainly located on the outer surface of a B-lymphocyte cell membrane, where it acts as a receptor for antigens.
- **IgE** is involved in the defence against parasitic infestations and allergic reactions.
- **IgG** is the major antibody involved in immune responses against any infections that have been previously encountered.
- **IgM** is the largest antibody, consisting of five antibodies bound together, and is the first type of antibody formed when the body encounters a new infection.

Secrets About Leucocytes and Antibodies

- Neutrophils make up the largest proportion (about 65 per cent) of the circulating leucocyte population. Eosinophils account for about 3 per cent, basophils represent about 1 per cent, monocytes make up approximately 5 per cent and lymphocytes account for 20–35 per cent of the circulating leucocytes.

- IgA makes up 15–20 per cent of the total Ab pool, IgD forms less than 1 per cent, IgE is only found in tiny amounts, IgG accounts for 70 to 75 per cent and, finally, IgM form 10 per cent of the total Ab pool.

- During exercise, your white blood cell (WBC) count rapidly increases so that your body can identify pathogens faster. The WBC count returns to normal when you rest. It may come as a shock to some people, but there are actually positive reasons to exercise regularly!

- Vitamin C every day helps keep the pathogens away! This is because it encourages WBC production to help fight off an infection. This means that citrus fruits, berries and cruciferous vegetables (i.e., broccoli and cauliflower) are a big plus in the fight against pathogens.

- Studies show that being overweight weakens your immune system and therefore your ability to fight off infection due to a lower WBC count. So shedding excess pounds will keep you healthier in the face of illness.

- If a person's white blood cell count is too high, it's not necessarily a good thing. It often signifies an underlying health problem – such as inflammation, trauma, allergy or other diseases such as Leukaemia.

A poor diet causes the WBC count to decrease, so it is vital to curb your fat, caloric, sugar and sodium intake in favour of foods that are high in antioxidants, fibre, calcium and healthy monounsaturated fats, including plant-based oils such as sunflower oil.

Platelets

These oval shaped cells are the smallest blood cell. Mature platelets do not have a nucleus, but they are made from much larger nucleated cells and found in the red bone marrow, lungs, spleen and liver.

Thrombocytes go through a dramatic change when they are activated in response to any tissue damage. Seconds after a cut to the skin, the muscular walls of blood vessels contract to reduce blood flow to the area, in order to minimize blood loss for up to 30 minutes! In addition to this vascular mechanism, platelets clump together and release their chemicals (enzymes), which start to make a blood clot within 30 seconds of the damage occurring. This clump (or plug) of cells and clotting (coagulation) process is completed in a few minutes to prevent even more blood loss. Platelets contain microfilaments, which, like muscles, contract, and this enables the clot to retract, helping the edges of the wound knit together to aid the healing process. Blood is converted quickly from its usual liquid state into an insoluble fibrous, gel-like red blood clot (since the clot traps red blood cells). The clot stops bleeding and acts as a physical barrier to any invading pathogens. These mechanisms are adequate in preventing blood loss if the damage is to small blood vessels, but if larger blood vessels are involved, resulting in huge blood loss, these mechanisms are inadequate, so you

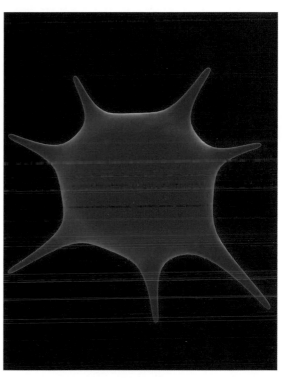

A platelet forms spines and spikes during the early part of the clotting process.

Haematology

The study of blood is called haematology and it involves estimating:
- **the number of cells and non-cellular blood components, and comparing these with normal ranges**
- **the shape and size of the cellular components, using a stained blood film.**

A full complete blood count screens for anaemia (lack of oxygen carrying capacity of blood), various infections and blood-clotting disorders. It includes counts of the red blood cells (RBCs), white blood cells (WBCs), platelets, a differential white cell (see later) and an estimate of the haemoglobin values.

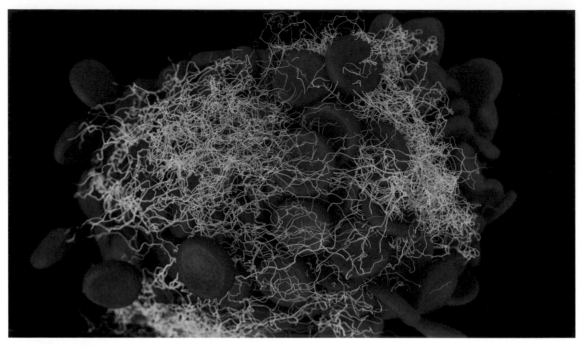

A clot of red blood cells trapped in a net of protein fibres. The fibres contract to pull the blood cells tighter together, hardening the clot.

need to seek medical attention to stop any bleeding and prevent significant and dangerous blood loss.

Clotting

Coagulation is a complicated pathway called the "clotting cascade". This cascade is a multiplicative process where inactive chemicals (enzymes) interact so that the conversion of one inactive enzyme produces an active enzyme that activates another inactive enzyme, and so on, in a chain reaction that increases in magnitude with each step. Many clotting factors from the plasma and platelets are involved.

Secrets About Blood

- The volume of blood circulating in an adult averages 5 litres (9 pints) in males and 4 litres (7 pints) in females.

- Your blood volume accounts for about 8 per cent of your total body weight.

- Plasma and red blood cells make up about 55 per cent and 44 per cent of a blood sample, respectively. The remaining 1 per cent consists of white blood cells and platelets.

- There are about 28,000 billion red cells in your blood circulation and you produce about 2 million new red cells every single second.

- Your blood normally contains 150,000–350,000 platelets per microlitre (3 fl oz).

- White blood cells have a wide range of lifespans, from hours to years. Platelets last between 9–12 days.

Making Blood Cells

Blood production, or haemopoiesis, in an unborn child occurs initially in the liver and spleen. As the unborn child grows and requires more blood, so the bone marrow takes over the production of all blood cells.

At birth, all the bone marrow is active in blood production, and it is called red active marrow because although it produces all blood cell types, the huge number of erythrocytes makes it red. As the infant grows, the active red marrow of the long bones of the limbs is replaced by yellow inactive, or fatty, marrow. In a young adult, the active marrow is found in the flat bones, such as the cranium, ribs and pelvis. The liver, spleen and the yellow inactive marrow can, however, revert back to erythrocyte production when red cell numbers are low, such as during severe anaemia.

Stem Cells

All types of blood cell are made from stem cells called haemocytoblasts. The production of red blood cells, called erythropoiesis, involves the development of immature erythrocytes whose nucleus and ribosomes control the production of the haemoglobin molecules. This process involves the cells' genes providing a template that codes for the haemoglobin, and the ribosomes making it by putting the necessary amino acids in the right order. Once the red blood cell is full of haemoglobin, the cell's nucleus, along with most of the organelles, are ejected, and the mature red blood cell enters the blood circulation. The rate at which red cells are produced is controlled by the hormone erythropoietin, which is produced in the kidneys. The kidney cells produce more erythropoietin if your blood is low in oxygen.

The bone marrow's other blood production activities are leucopoiesis (white cell production) and thrombopoiesis (platelet production).

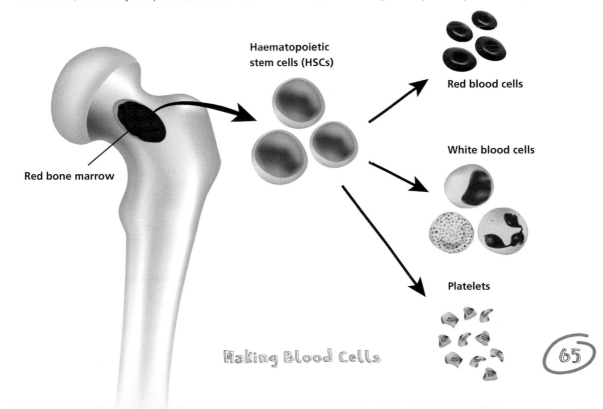

Red bone marrow

Haematopoietic stem cells (HSCs)

Red blood cells

White blood cells

Platelets

Bone Marrow Biopsy

A bone marrow biopsy provides information about the status of blood cell production, the number of the blood cells and the presence of abnormal cells, as found in cancers. The procedure involves puncturing the bone with a needle, then taking a specimen of the bone marrow with a syringe. The sternum is a common site of biopsy in adults, and the tibia (shin bone) is a common site in children, since these areas have relatively thin layers of bone above the marrow, making it easier to puncture.

Blood Doping

Some athletes have used blood doping and the administration of erythropoietin in an attempt to improve their performance, particularly during endurance events. This involves removing blood cells from the body, storing them for four to five weeks, then reintroducing them into the body a couple of days before an event. This increases the oxygen-carrying capacity of blood. Blood doping is banned by the International Olympics Committee because of the extra workload it enforces on the heart due to the increased viscosity of blood as a result of the addition of extra erythrocytes.

How Do Microbes Spread?

There are many ways that dangerous microbes can enter the body. These include:

- **Person-to-person** – by the mixing of blood, when the skin epidermis has been broken, for example by sharing needles; saliva, by kissing; or, for very infectious diseases, through the air, via coughing or sneezing.
- **By food** – bacteria may survive in food if it is inadequately cooked or if it is reheated.
- **By water** – contaminated water may spread diseases, such as cholera.
- **By insects** – for example, malaria is carried by mosquitoes in certain parts of the world.

Once inside the body, these microbes need to be dealt with by our leucocytes (sometimes called immune cells) circulating in blood and the lymphatic system to prevent their spread causing any diseases. Usually signs and symptoms of infection or disease are not obvious and this is because our normal (basal) level of already circulating leucocytes and their products, such as antibodies, are enough to deal with lots of microbes and abnormal cells. However, if the number of these invaders exceeds the body's ability to fight them off, then signs and symptoms of an infection or disease soon become apparent, unless we react fast.

The body reacts by increasing the number of troops and weapons, called leucocytes and antibodies, respectively, to fight off these invaders. A differential white blood count may determine the invading organism or abnormal cells. For example, a raised neutrophil count may indicate a bacterial infection and cancer bio-markers may indicate the presence of a tumour.

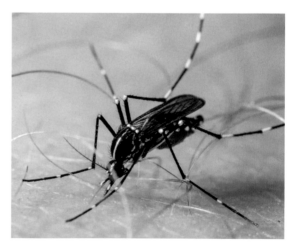

Mosquitoes are responsible for the spread of malaria, which caused 429,000 deaths in 2015 according to the World Health Organization.

Lymphatic and Immune Systems

The lymphatic system is an extensive, branched network of tubes that is aligned closely with the blood's circulation. One of its key roles is to prevent the build up of too much fluid in the tissues.

Blood distributes oxygen, nutrients, hormones and other substances to the body's cells, and removes waste products, such as carbon dioxide. Cells are bathed in tissue (interstitial) fluid, which acts as a link between the blood and the cells. Excess tissue fluid is returned back to the circulation, so the lymphatic system is important in the homeostatic maintenance of all body fluids. Swollen lymph nodes are a sign that this filtering activity is compromised and it may be that an infection is

The Lymphatic System

Tonsils – two lymph-nodes at the back of the throat.

Thymus – a lymph gland found between the trachea (windpipe) and the sternum (breastbone).

Spleen – a lymph organ in the upper-left side of the abdomen positioned just behind and above the stomach.

Lymph node – a filtering area within the lymphatic vessels where lymphatic cells are concentrated.

Lymphatic vessel – a network that channels tissue fluid back to the veins, filtering out potential sources of infection with its rich supply of lymphatic cells.

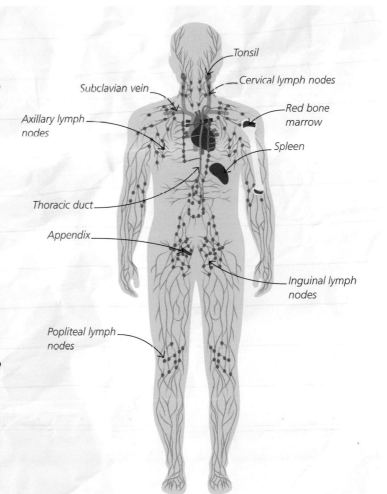

Tonsil

Cervical lymph nodes

Subclavian vein

Red bone marrow

Axillary lymph nodes

Spleen

Thoracic duct

Appendix

Inguinal lymph nodes

Popliteal lymph nodes

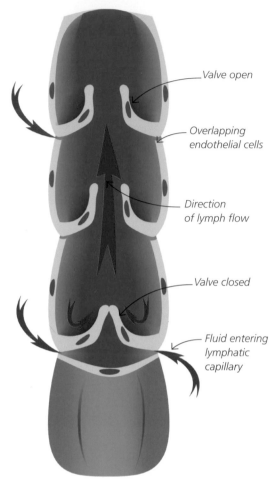

Valve open

Overlapping endothelial cells

Direction of lymph flow

Valve closed

Fluid entering lymphatic capillary

The valves in a lymphatic vessel prevent lymph from flowing back towards the body's interstitial fluids.

smooth muscle cells in the walls of large lymph vessels also pushes fluid along until it terminates into two large lymph ducts that empty lymph into blood vessels in the neck region.

Lymph Node

On the way from the beginning of each lymph vessel to their termination at the blood vessels in the neck, lymph vessels divide to form incoming (afferent) lymph vessels that lead into oval-shaped, encapsulated lymph nodes. Nodes exist individually, in chains along the lymphatic system, or randomly, such as those in the respiratory mucous membranes. They can also exist as multiple groups that are found at specific sites, such as the tonsils. Each node has internal extensions of the capsule, and these house the germinal centres where many macrophages and lymphocytes accumulate. Any antigenic material present in the lymph is filtered out by the sieve-like action of the lymph node and immediately attacked and hopefully destroyed by the macrophages and lymphocytes. Filtered lymph leaves by an outgoing (efferent) lymph vessel and flows into the two lymphatic ducts in the neck.

present, since the lymphatic system has immunological defence functions, filtering and destroying potential environmental hazards, including antigens.

Lymph vessels are closed at one end with a special valve-like junction. This allows lymph to enter the vessel, but not flow back into the tissue fluid. Lymph capillaries drain lymph fluid (i.e drained tissue fluid) into larger vessels. These have special valves that promote one-way flow, just like vein valves, so lymph only can be returned to the blood. Lymph is massaged through the lymph vessels by contraction of the skeletal muscles during body movements. Rhythmic contraction of

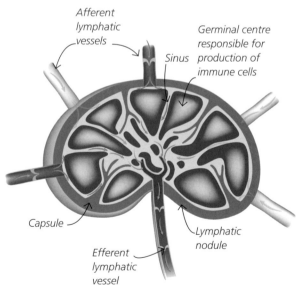

Afferent lymphatic vessels

Sinus

Germinal centre responsible for production of immune cells

Capsule

Efferent lymphatic vessel

Lymphatic nodule

Lymph nodes act as filtering units in the lymphatic system, cleaning the lymph of antigens before it flows back into the blood.

Can I Live Without a Spleen?

Because of its soft consistency, rupturing the spleen is quite common in traumatic injuries, such as broken ribs. Any rupture causes severe bleeding (haemorrhage), and may lead to shock (a sudden drop of blood pressure). Following diagnosis, the condition can be stabilized with a blood transfusion. The removal of the spleen (splenectomy) is performed to prevent the patient bleeding to death. Once bleeding has been controlled, the person soon recovers.

A missing or non-functional spleen means that people may be more prone to microbial infection, and therefore special immunization programmes are recommended to the patient.

People can live without a spleen since blood cells, especially white cells, produced in other areas of the body, such as bone marrow and the liver, take over. There is also a mass of other lymphatic tissue to compensate for the spleen's former immunological activity.

The Spleen

The spleen mimics a large lymph node as it houses an army of macrophages and lymphocytes. This organ is situated in the upper left region of the abdomen, lying between the stomach and the diaphragm. Blood vessels and nerves enter and leave the spleen. It also has an outgoing (efferent) lymphatic vessel, but it does not have an incoming (afferent) lymphatic vessel. As a result, it filters arterial blood of antigens, but it cannot be exposed to infections spread by the lymphatic system.

The spleen also destroys erythrocytes when they have come to the end of their life. The products of haemoglobin breakdown are released from the spleen into blood and taken to the liver and the bone marrow where they are recycled into new red blood cells. Leucocytes, platelets and microbes are also destroyed in the spleen. Finally, the spleen acts as a blood reservoir, releasing blood on demand following severe blood loss, so helping to maintain the composition of body fluids.

The Thymus Gland

The thymus is a major organ of the lymphatic system and is located in the upper chest, just above your heart and just behind the sternum. It is large at birth and continues to grow slightly into puberty. During adolescence, however, it starts to reduce its size, and by middle age it has returned to its size at birth. Despite this size reduction, the thymus continues to function throughout life. However, the effectiveness of its lymphocytes in response to antigens declines.

The thymus is primarily responsible for the production of T-cells. T-lymphocyte stem cells are made in the bone marrow, where stem cells differentiate into the specialized T-lymphocyte. The majority of these cells enter the thymus, where they undergo cell division (mitosis) and mature into adult T-cells. These mature T-lymphocytes cells enter systemic blood, and are transported to lymphoidal tissue, or they remain in the thymus, to become future generations of T-lymphocytes.

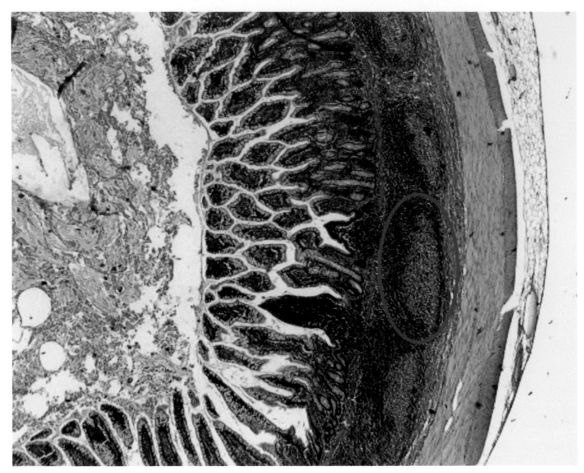

A Peyer's patch (ringed in red) found in the wall of the small intestine. There are about 100 of these patches in the human intestine.

Peyer's Patches

The human gut is one of the many areas where antigenic insults, which may be present in food, enter the body. To counteract this and protect the body, the wall of the intestine is lined with immune cell lymphoid tissue, which is also clustered in the tonsils, adenoids and the appendix. In addition to this, the Peyer's Patches are lymphoid nodes, located mainly in the small intestine, but they are also found in parts of the large intestine. These patches have an oval or round shape and they offer a protective defence against pathogens and their toxic chemicals contained in the food we eat. They also protect against pathogens that have gained entrance to the mouth and anal canals. These patches have the ability to detect directly within the gut any pathogens, such as salmonella, that have an affinity for this region. These cells trigger off the complex set of cellular and chemical reactions associated with immunity to destroy the invaders and any harmful chemicals associated with them. The patches, for some reason, are more numerous in younger people and become less prominent with age, although doctors don't know the reason for this decline. However, this may be one of the reasons why the elderly are more susceptible to gut-related infections!

Internal Protectors: Warriors Behind the Scenes

How Diseases Spread

Lymph capillaries drain fluid from the tissues, and this may contain pathogens and tumour cells. If these are not destroyed by your immune cells, they may settle and multiply in the first lymph node they encounter, producing a localized infection, such as tonsillitis, or even tumours.

Alternatively, following multiplication, these cells may spread to other lymph nodes, blood or other parts of the body, using the body's transporting systems. As a result, each new site of infection or secondary (metastatic) tumour becomes a further source of infection or cancers using the same routes. These can then produce infections or tumours elsewhere, such as glandular fever, or thymoma (cancer of the thymus gland). Breast cancer frequently shows lymphatic spread.

Infections and tumours can cause enlargement of lymph nodes and lymph organs. With a "sore throat" or a "cold", the cervical (neck) lymph nodes enlarge in response to the infection. Most people often claim that their "glands are up", but the lymph nodes are not glands, as they do not release secretions. Glands are "up" because the nodes are actively producing lymphocytes to defend against antigens.

Your body's immune cells multiply and mobilize in response to an infection.

Protecting the Body

Your body's immune system is made up of the mobile army of white blood cells that patrol your body to protect you against infection and disease. The armed forces are primed to identify, destroy and kill organisms and substances (antigenic insults) that are not usually found within your body. These include:

- **Disease-causing microbes** called pathogens, such as bacteria, viruses and fungi.
- **Poisonous substances** released from pathogens, for example, bacterial toxins.
- **Infected or diseased cells**, such as cancer cells, which tend to make abnormal proteins.
- **Transplanted tissues**, such as unmatched blood.
- **Environmental pollutants**, for example car exhaust fumes.
- **Foreign bodies**, such as wood splinters and gunshot shrapnel.

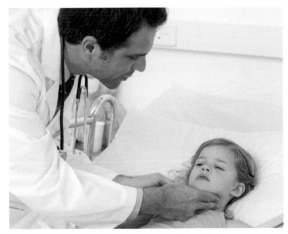

Lymph nodes in the neck, or glands, only become swollen and painful when a microbe has infected them.

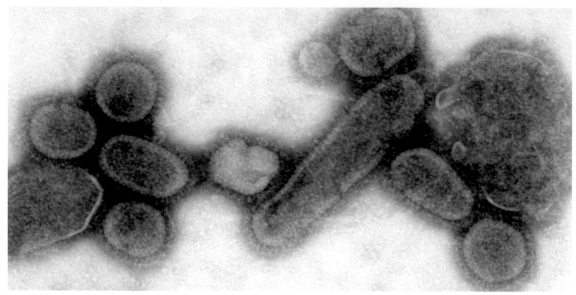

This grainy photo shows the 1918 Influenza virus, responsible for one of the deadliest pandemics ever recorded. It infected some 500 million people and caused 50–100 million deaths.

Main Defences

There are two main lines of defence. The first line or non-specific immune response, is programmed into everybody at birth. It includes mechanisms that provide general protection against the invasion of a wide range of pathogens. These mechanisms are the same for everyone, hence the term "non-specific". They include the external physical barriers such as the skin's layered epidermis, and chemical barriers, such as the secretions of the skin's epidermis – sweat and sebum. They also include the gastrointestinal and respiratory tracts, and internal (bodily) reactions, including the phagocytic response, which provides surveillance against microbes and their toxins.

The second line, or acquired immunity, develops as your body's mobile army of warriors (leucocytes) come up against antigenic insults. This immunity develops slowly during your lifetime. Memory cells form that are primed to recognize foreign proteins. Dormant (inactive) memory cells patrol your body and are activated only if they encounter the antigenic insult again. The body system responsible for acquired immunity is the lymphatic system, as it involves interaction between the two types of leucocytes – B-lymphocytes, which produce antibodies, and T-lymphocytes, which regulate the production of antibodies and of other chemicals which attack the foreign proteins.

Both non-specific and specific immunities are adapted to maintain the equilibrium (homeostasis) of the body's internal environment.

Immunity in Young and Older People

Newborn babies are more susceptible to infection and disease, since they need to develop their specific immunological responses by exposure to environmental antigenic insults. The ageing process is associated with the increased destruction and decreased production rates of leucocytes. It is, therefore, not surprising that elderly people are also more susceptible to diseases. A lack of gastric juice is also more common in old people, which potentially makes them more susceptible to pathogens in their diet.

Know Thyself!

As well as recognizing antigenic insults, your immune cells must learn to recognize normal components of your body and leave them alone, but also identify body cells that have changed, as a result of infection or disease, since these need to be destroyed.

A cell contains a number of "self" or "identity" tags on its membrane that brand it as part of the cell. These "self tags", known as Human Leucocyte Antigens (HLAs), are coded by a group of genes known as the major histocompatibility complex. The immune cells learn to recognize these markers during the development of the embryo and do not attack body cells known as "self".

Body cells repeatedly break down proteins within the cell, and the byproducts of this are transported to the cell membrane and displayed on it, along with the HLAs proteins. As a result, each cell shares information about its internal conditions to immune cells. If the immune cells encounter a virally infected cell or diseased cell, such as a cancer cell, with both self tags (HLAs) and foreign proteins (antigenic insult), then the cell is usually identified as undesirable and attacked and destroyed.

Hospital-Acquired Infection

At the turn of the century, of all the patients entering hospital without an infection, 10 per cent would contract a hospital-acquired infection, and at any one time about 20 per cent of all patients in hospital were suffering from a hospital-acquired infection. In the United States alone, the Centers for Disease Control and Prevention estimated that there were about 1.7 million hospital acquired infections. In these environments, infections can be spread by health care staff, bed linen, contaminated equipment and air droplets, or it can come outside factors, including other patients and visitors.

Today, even with all the advances in hygiene, a patient may still be at risk of acquiring the much-publicised MRSA and Clostridium infections.

This image shows a white blood cell (green) attacking purple MRSA bacteria.

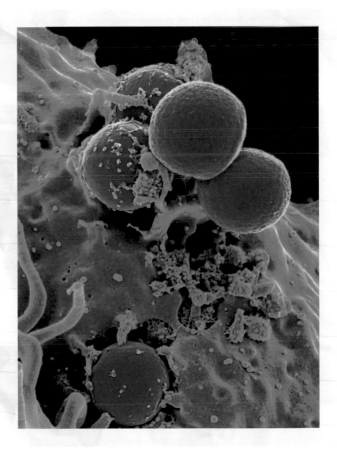

Immunization Against Diseases

The immune system can be encouraged to produce antibodies by immunization. When you are immunized, your blood is injected with antigenic insults, which include dead or inactivated (weakened) pathogens. These include bacteria or viruses, or harmless proteins (toxoids) that have been removed from pathogens. These vaccines retain antigenic properties and your B-lymphocytes are stimulated to make antibodies against the injected antigenic insult. These form a group of B-memory cells that will respond to those particular antigenic insults. If that particular virus or bacteria comes into the body again, the immune system is able to identify and destroy it speedily before it causes a serious illness. However, it can take weeks for the antibodies to reach an adequate protective level. This is known as an active immunization because you are being stimulated to actively produce your own antibodies against specific pathogenic antigenic insults.

Alternatively, passive immunization involves injecting ready-made antibodies from human or animal immune donors. These provide rapid

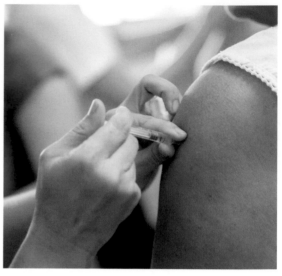

Many countries organize nationwide vaccination programmes to protect against some diseases.

protection for non-immune people, for example health care professionals exposed to dangerous infections, such as hepatitis. The level of protection slowly declines after donated antibodies are naturally removed from the recipient's body.

How to Save a Nation!

In 1796, Edward Jenner used the fluid extracted from cowpox blisters to provide immunity against the similar virus of smallpox. Modern immunization programmes have reduced the incidence of many serious diseases, including whooping cough, tuberculosis, measles, diphtheria, cholera, rubella, smallpox, typhoid and poliomyelitis.

The success of these programmes means that some diseases, such as polio, no longer occur in many countries, and smallpox has been eradicated globally! In other countries where immunization is not so widely available, this is not the case. Since overseas travel is so popular now, there is a risk that diseases could be brought back into a country and spread to people who have not been immunized against these diseases.

Chapter 5
THE RESPIRATORY SYSTEM

Respiration

Respiration is a scientific term that refers to the body's use of oxygen to release energy in the form of a chemical called adenosine triphosphate (ATP), which is readily accessible for cell activities.

Although, the production of most ATP involves oxygen, called "aerobic" or mitochondrial respiration, some ATP is generated in cells in the absence of oxygen, known as "anaerobic" or cytoplasmic respiration. The oxygen requirement of individual cells differs according to the amount of energy they need. For example, extremely active cells, such as your heart muscle cells, use about 25 to 35 times as much oxygen per minute than your relatively inactive skin cells. As a result, active cells' cytoplasm is packed with mitochondria – "the secret powerhouse of the cell". Your oxygen requirements also respond to different situations, for example, your skeletal muscle cells' oxygen usage for ATP production may increase 20 times during exercise so that the ATP can provide enough energy for continued muscle activity.

Keeping a Balance

Maintaining a sufficient supply of oxygen is therefore a priority of your body in maintaining an equilibrium within cells and tissues. Tissue oxygenation takes place through the following stages: oxygen is taken from the outside air into your blood through the respiratory system; oxygen is then transported by your blood circulation to the tissue cells; and oxygen then passes from the blood to your cells, where it can be used in cellular respiration in the manufacture of ATP.

In cells, the oxidation of the chief "fuel" – glucose – produces ATP, and also carbon dioxide and water. These play a key role in determining the "acidity" of fluid within your cells, tissue fluid and blood. If you produce too much carbon dioxide and water, this may cause a potentially dangerous situation called an acidosis, in which enzyme activity, and your health, may be compromised. In order to counteract this potential condition, excess carbon dioxide is removed from the body through the lungs. The activity of the lungs can therefore be summarized as exchanging oxygen for carbon dioxide, and the control of breathing relates to the body's needs in this respect. Carbon dioxide removal from the cells, blood and body, via the respiratory system, involves the opposite sequence of events as those identified above for oxygen.

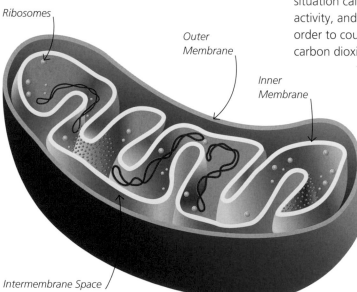

Ribosomes

Outer Membrane

Inner Membrane

Intermembrane Space

Mitochondria are the key structures involved in respiration inside the body's cells.

Airways and Breathing

The respiratory tract consists of the passageways that carry air to and from the gaseous exchange surfaces of the lungs. It is divided into two main parts – the upper and lower respiratory tracts.

The upper respiratory tract is made up of the nose (really the nostrils), nasal cavity, sinuses and the throat (or pharynx). As air is breathed in and passes through these passageways it is filtered, warmed and moisturized or humidified, while air that is breathed out is cooled and dehumidified. Most people breathe through the nostrils of the nose, although some people breathe through their mouths, particularly if the body's oxygen requirements are higher, such as during strenuous exercise.

The lower respiratory tract consists of the windpipe (trachea), voice box (larynx), two bronchi, bronchioles and air sacs (alveoli) of the lung where all gas exchange occurs between blood and air.

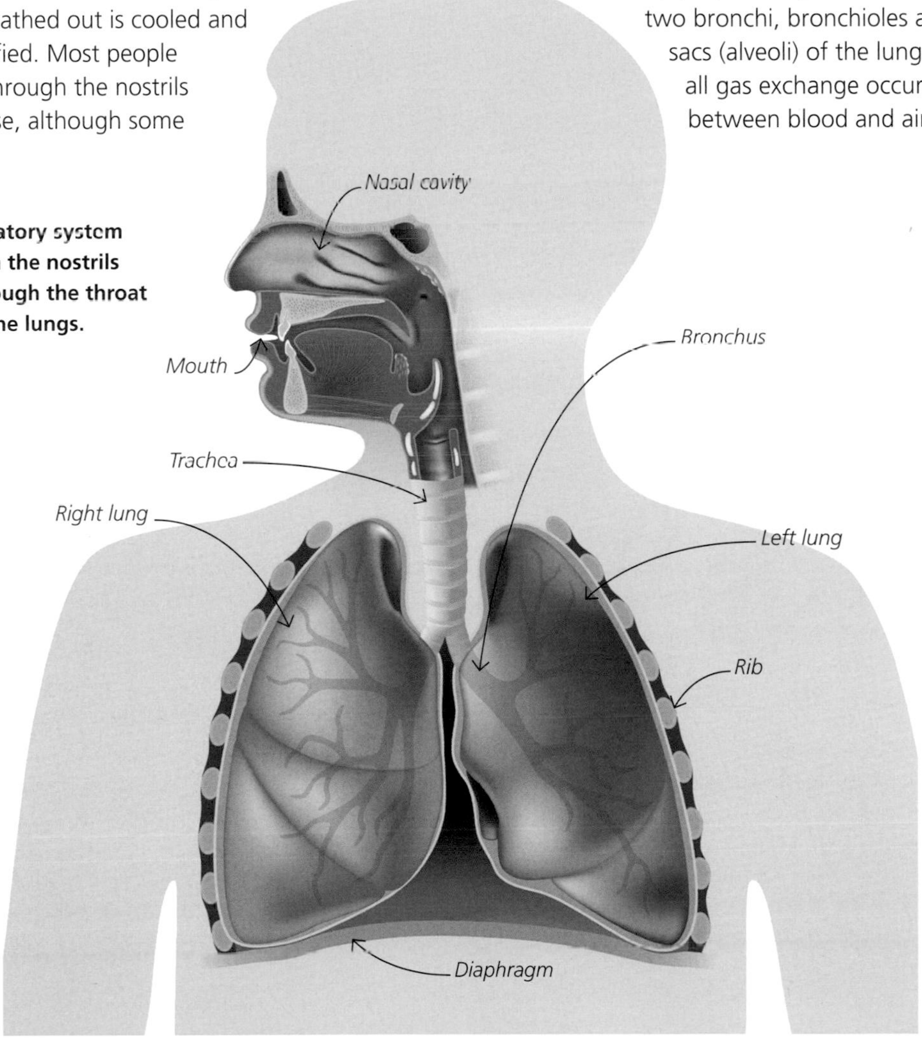

The respiratory system leads from the nostrils down through the throat and into the lungs.

Nasal cavity

Mouth

Bronchus

Trachea

Right lung

Left lung

Rib

Diaphragm

Cilia

The lining of the airways are covered with a dense carpet of hair-like structures called cilia. Surrounding the cilia are goblet cells, which produce mucus to stop the airways from drying out and to collect dust particles and any microbes we may have breathed in.

Cilia are sometimes called "motile cilia" because they beat in a co-ordinated fashion producing waves of movement along the airways. This movement resembles blustery wind moving wheat in field. Cilia activity in the airways is designed to move mucus and any trapped dust and microbes we breathe in towards the larynx and pharynx, where they are swallowed. Once in the stomach, these microbes are then destroyed by the stomach's acidic, enzyme-containing juice, or they are coughed up as phlegm.

Some chemicals inhaled from cigarette smoking slow down the active movement of cilia and researchers have suggested that, over a period of

Cilia act as street cleaners, sweeping your airways clear of debris.

time, this may be responsible for clogging-up of the airways, and the development of the "morning cough" that seems to occur more frequently in smokers than non-smokers. Only time will tell if electronic cigarettes will have the same effect on the airways!

Are You a Mouth Breather?

Mouth breathing happens for a number of reasons, but the usual cause is a nasal obstruction. If you have problems breathing through the nose, you are forced to take in cold, dry air through the mouth. Other reasons for mouth breathing are: your bite might be slightly off; your jaw may be incorrectly aligned when you are asleep, so your mouth is not closed. Kids may have abnormally large tonsils, they may have a birth defect, like a deviated nasal septum that may make it more difficult for them to breathe through their nose. It could even be a skeletal problem, which causes someone to lean forwards and breathe through their mouth. Whatever the cause, you should try and avoid breathing through your mouth, to prevent its side-effects.

A common side-effect is an extremely dry mouth, which can be a breeding ground for bacteria which may give rise to further problems like bad breath, bleeding gums and cavities.

In children, mouth breathing can also lead to skeletal issues, since it encourages growth of the upper jaw and can cause a large overbite and "gummy smile". Mouth breathers frequently wake in the night, because they are not getting as much oxygen, which can have a knock-on effect on your energy levels, attention and concentration, thus affecting your daily life.

Natural mouth breathers may be able to "stave off" dryness by hydrating the mouth throughout the day. However, the mouth will get dry overnight because they are breathing through their mouth all night, and this causes the soft tissues in the mouth to dry out.

If you're a mouth breather, but don't want to be, then it is important to determine the cause before you can correct it. If the problem is huge tonsils, then removing them might be an option. Using a humidifier while sleeping may ease mouth dryness, as can drinking lots of fluids.

Respiratory Reflexes

Breathing is controlled by the respiratory centres in the brain. Sensory information is delivered by receptors from all over the body and these centres can then decide if a change in respiration rate is needed in order to maintain blood gas homeostasis.

The blast of air in a sneeze forces any irritants out of the nasal cavity – as well as a lot of mucus!

The body's sensors collect information on a wide range of attributes. These include the levels of respiratory gases (carbon dioxide and oxygen), acidity levels in the blood and cerebrospinal fluid, changes in blood pressure, the amount of stretch in lung tissue, and other sensations such as nasal irritation. This sensory input alerts the breathing pattern to either increase or decrease the rate and depth of breathing. Specific respiratory reflexes are important, in order to prevent the overinflation of your lungs, or reduce your exposure to respiratory irritants, such as strong perfumes and deodorants.

Sneezing

Sneezing helps to clear the upper respiratory passageways when the nasal cavity lining is irritated by inhaled particles, strong odours or by the presence of infections. When you sneeze, it begins with one or a few large inhalations ("Ah... Ah... Ah"). The sneezing reflex closes your glottis (and your eyes), and then a powerful contraction of your respiratory muscles opens your glottis and forces air out through your nose and mouth producing the characteristic sound of "Ah....chooo"!

Sputum in Health and Infection

Sputum is a mixture of mucus and saliva that is removed from the airways by coughing. In health, it is colourless and of a slightly thick consistency. However, if you have an infection, the mucus part of sputum may contain pus (dead microbes and dead body cells). The amount, colour, consistency and odour of sputum varies with different pathogens. For example, the presence of the microbe *Pseudomonas aeruginosa* causes a greenish sputum, whereas the viral microbe that causes pneumonia produces a small amount of sputum, and bacterial pneumonia produces much more sputum and is seen to be "more productive". A pear-drop (acetone) smelling breath is an indicator that the person may have diabetes. Other clinical investigations, however, must confirm that diagnosis.

- Half a litre (almost a pint) of water is lost from the body through your lungs every day.

- The common cold is by far the most frequent upper respiratory tract illness known today.

- There are more than 200 different viruses that can cause the common cold.

- Inflammation of the lining the upper respiratory tree system, caused by infection or allergies, causes increased production of mucus.

- During cold weather, the cilia of your respiratory system work more slowly, so that your nose starts to "run" and may even drip if you are not quick enough to catch it with your tissue.

Coughing

Coughing helps to clear your trachea and bronchi when cough receptors are irritated by inhaled particles or excessive mucus. A cough starts with a short inspiration, followed by a series of forced expirations against a closed glottis. The diaphragm and the other breathing muscles then forcefully contract, causing a sudden "snap" opening of the glottis. This produces an explosive outflow of air along with the nasty irritants.

Yawning and Hiccups

Yawning is an involuntary respiratory reflex learnt in the unborn child. It is believed to be an involuntary inspiration triggered by raised blood levels of carbon dioxide, to draw more oxygen into the lungs and raise blood oxygen levels. It is also thought to cool the blood as it passes through the brain.

Hiccups are actually activated by nerves in the diaphragm, which cause sudden involuntary recurring contractions of the diaphragm with the glottis closed.

Yawning may be a sign that your body is not getting enough oxygen to its cells, but scientists don't know why yawns can be infectious.

The Lower Respiratory Tract

The largest lower respiratory airway is the trachea. This tube receives air from the larynx, before it divides into two smaller airways – the left and right bronchi.

Each bronchus enters a lung, where it divides, like the roots of a tree, into smaller and smaller bronchi, which further branch into the bronchioles. These small airways branch again and again, until they form the terminal bronchioles, which open into clusters of tiny cup-shaped sacs called alveoli.

Each alveolus has a fantastically rich capillary blood supply. Being cup-shaped and numerous (there are about 350 million in both lungs!), alveoli provide a massive surface area for exchange of the respiratory gases – their total area is estimated to be about 70 square metres (230 sq ft).

Air in the Lungs

Since only your alveoli form the surfaces within the lung for the exchange of the respiratory gases, the rest of the airways (the nasal cavity, pharynx, trachea, bronchi and bronchioles) do not take part in the uptake of oxygen or the removal of carbon dioxide from the blood, so these airways are said to

Inside the lungs, the airways branch into smaller and smaller tubes before ending at the tiny alveoli. Each alveolus measures about 0.3 mm ($^1/_{100}$ in) in diameter.

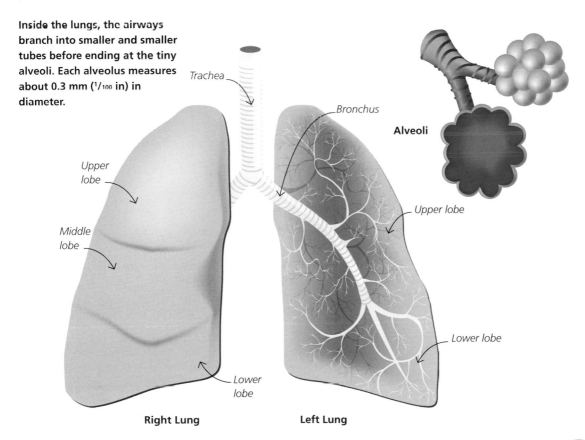

Trachea

Bronchus

Alveoli

Upper lobe

Upper lobe

Middle lobe

Lower lobe

Lower lobe

Right Lung

Left Lung

comprise a "dead space". However, these airways will be filled with air during breathing in, so only a proportion of the air we breathe in actually enters the alveoli. Furthermore, the lungs never deflate completely when we breathe out (even when we exhale as hard as we can) and so the alveoli and airways are filled with gas left over from the previous breath. Thus, when we breathe in, the first "air" to enter the alveoli is that left in the airways, and so the proportion of fresh air that enters the alveoli when we breathe in (inhale) also has to mix with that gas. This seems inefficient, but it is in fact helpful, as the mixing of air with alveolar

Blowing into the mouthpiece on the left of this peak flow meter will move the red indicator to show a person's peak expiratory flow, allowing them to monitor the potential effects of asthma.

gas means that the gas composition is only slightly enriched with oxygen and depleted of carbon dioxide, and so dramatic changes in gas composition of the blood leaving the lungs are avoided.

Secrets About Breathing

- **How quick you breathe depends on your developmental age. When you were born, your breathing rate is 40 to 45 times per minute. As you reached young childhood, your rate will have been reduced to about 30 times per minute, and in late childhood, it decreases even more to about 20 times per minute! By the time you are an adult, your rate will have been reduced to around 12 to 15 times per minute at rest.**

- **Adults breathe between 20 to 45 times per minute when they are exercising, unless they are a super athlete, who can breathe at approximately 60 breaths per minute during strenuous exercise.**

- **The average time an adult can hold their breath for is between 30 and 60 seconds.**

- **A cough forces air out of the lungs at speeds between 1.5–3 m/s (5–10 ft/s), whereas a sneeze forces air out at about 2.7 km/s (1.7 miles/s).**

- **Sneezing is regarded by researchers as a double reflex, since it is impossible to sneeze with your eyes open.**

Free divers try to dive as deep as they can or swim as far as they can on a single breath. Denmark's Stig Steverinsen currently holds a Guinness World Record for the longest free dive in 2010. He held his breath underwater (without using scuba diving equipment) for 22 minutes.

Gas Exchange

The air we breathe contains much more oxygen than the amount we need to live. In fact, the air that we breathe in consists mainly of nitrogen gas. Atmospheric air contains about 21 per cent oxygen and your body removes just a small percentage of this.

The air you breathe out contains about 16 per cent oxygen (don't worry, this is enough to provide ample oxygen during mouth-to-mouth resuscitation). Air that we breathe in also contains trace amounts of carbon dioxide (about 0.04 per cent), but the air we breathe out, contains 100 times this amount (about 4 per cent). This is because carbon dioxide is one of the end products of respiration. This gas may play an important role in controlling the acidity of body fluids, but it must be removed before it builds up to toxic levels. Uptake of oxygen into blood, and removal of carbon dioxide from it, takes place by diffusion – gases move from areas of high concentration to areas of low concentration – and so is based on the gradients in the alveoli and in the blood.

Under Attack!

The air we breathe in also contains invisible dust particles and microbes, and these can pose a threat to our health. Specialized phagocytic white blood cells are located in the alveoli and they "digest" these antigenic threats. Frequently, these white blood cells are seen to contain granules of atmospheric particles, such as black granules of carbon, that they have picked up from the alveoli surfaces inside the lungs.

These carbon deposits are commonly found in smokers and people who work with coal. Other white blood cells (B-lymphocytes) produce antibodies, which together with the macrophages, make it difficult for microbes to enter the blood and tissues.

The rich blood supply flowing around the alveoli ensures that plenty of oxygen diffuses into the blood, while carbon dioxide diffuses the other way. Air and blood never come into direct contact, so the gases have to pass through the cells forming the walls of the alveoli and the blood capillaries.

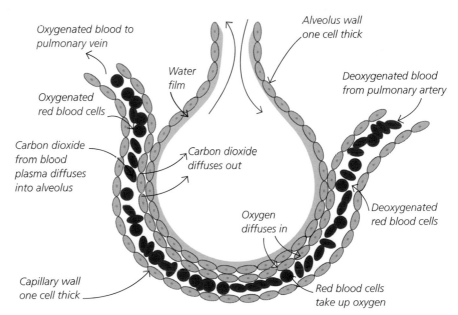

Oxygenated blood to pulmonary vein

Alveolus wall one cell thick

Water film

Deoxygenated blood from pulmonary artery

Oxygenated red blood cells

Carbon dioxide from blood plasma diffuses into alveolus

Carbon dioxide diffuses out

Oxygen diffuses in

Deoxygenated red blood cells

Capillary wall one cell thick

Red blood cells take up oxygen

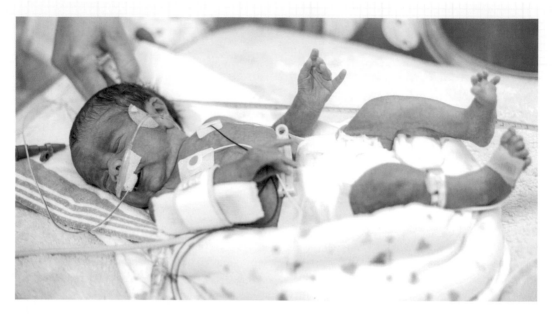

Surfactant and Premature Babies

Surfactant chemicals are composed of lipids and proteins, which are produced by specialized cells in the lining of the alveoli. Surfactant prevents the alveolar wall from collapsing. It is not produced in quantity until approximately the 34th week of pregnancy and after that there is a surge in its production. A lack of surfactant in premature babies can result in a reduced alveoli surface for respiratory gas exchange in the lungs and, therefore, its lack is a major contributory factor to whether they survive or not.

The chances of survival are increased if the alveoli can be opened more effectively. Researchers are currently developing artificial surfactants that can be delivered by inhalation, and these surfactant chemicals will make sure that the alveoli do not collapse, but remain open until the baby can produce sufficient amounts of surfactant on their own. If premature delivery is thought to be likely, then the health care professional looking after the expectant mother can administer dexamethasone (a steroid drug) as this has been found to accelerate the production of surfactant.

Lungs

The lungs are paired cone-shaped organs inside the chest cavity. The lungs are separated from the abdomen by a large sheet of muscle called the diaphragm. This muscle is dome-shaped prior to lung expansion, but flattens during breathing in. These actions of the diaphragm are essential to lung "inflation" and "deflation". Inside the lungs are the branching airways that actually begin with the nasal cavity, but extend to the minute gas-exchanging alveoli.

Each lung is surrounded by a pleural sac, which has two layers. The inner visceral pleural membrane covers the surface of the lung, while the outer parietal pleural membrane is attached to the chest wall. In between these two membranes, there is about 5 ml of lubricating (pleural) fluid that allows the lungs to slide over the chest wall when they inflate (or expand) as air is drawn into them during breathing in (inspiration), and deflate when air is removed from the lungs when you are breathing out (expiration).

The Respiratory System

Aspiration

"Aspiration" is a term used when objects or fluids penetrate into the airways when you breath in. Tiny particles will normally make you cough or sneeze, and the force of exhalation accompanying these reflexes is usually enough to remove the objects. However, medium-sized particles may remain in your trachea or even travel down to the primary or secondary bronchi, in which case they are usually stuck in the right lung. This is because the left-sided positioning of the heart means that the right bronchus is larger and more vertical than the left one. Striking someone on the upper back or performing the Heimlich manoeuvre may even be

The Heimlich manoeuvre involves standing behind a choking person and squeezing with your arms below their rib cage and up into their diaphragm. This forces air out of the lungs, taking any blockage with it.

Secrets of the Lower Respiratory Airways

- **The larynx is made up of cartilage and ligaments that form the "Adam's apple". The Adam's apple is found in both sexes, but is much more prominent in men.**

- **The male sex steroid hormone, testosterone, is responsible for causing the enlargement of the larynx and the deeper voice tone that accompanies it.**

- **The trachea is strengthened by 16 to 20 C-shaped rings of cartilage to stop it from collapsing.**

- **The left lung has two lobes, while the right has three. The left lung is smaller than the right because of the space occupied by the heart.**

- **Lower respiratory tract infection (LTRI) is a general term used for an acute infection of the trachea airways branching from the trachea and the lungs, which make up the lower respiratory system. LTRIs include bronchitis and pneumonia.**

- **It is possible for you to live with just one lung! However, you will need to restrict your physical ability, and adapt your life accordingly.**

necessary, since the coughing reflex may not move such a large sized object.

Aspiration of larger-sized particles and fluids normally does not occur and therefore is even more dangerous. The risk is magnified further if you are experiencing an altered state of consciousness, or when neural reflexes are not working as they should, such as if you are a drug abuser, or you have undergone trauma to the muscles associated with breathing. Larger objects are usually prevented by the larynx from passing deeper into the lungs. On the other hand, fluids may pass deep into the lungs, should aspiration occur, and these cause severe inflammation. An example could be the aspiration of stomach acid which has refluxed out of the stomach.

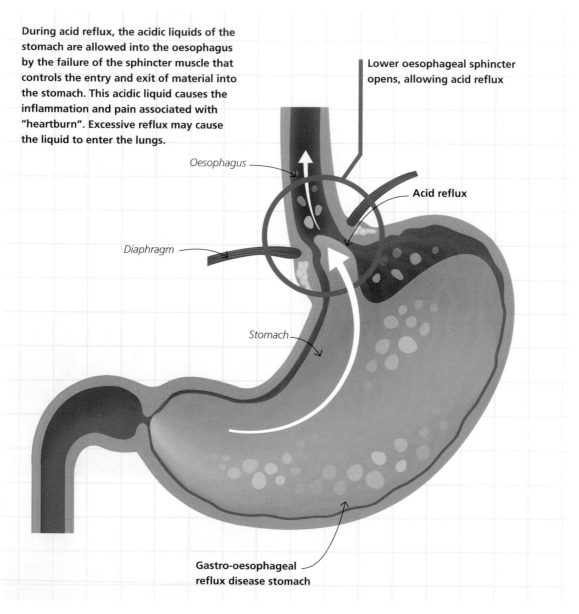

During acid reflux, the acidic liquids of the stomach are allowed into the oesophagus by the failure of the sphincter muscle that controls the entry and exit of material into the stomach. This acidic liquid causes the inflammation and pain associated with "heartburn". Excessive reflux may cause the liquid to enter the lungs.

Lower oesophageal sphincter opens, allowing acid reflux

Oesophagus

Acid reflux

Diaphragm

Stomach

Gastro-oesophageal reflux disease stomach

The Respiratory System

Transporting Gases

The transport of respiratory gases by blood, especially that of oxygen, is made easier by the red pigment haemoglobin, found in erythrocytes. In each of your red blood cells, there are 200–300 million haemoglobin molecules.

Each haemoglobin molecule is made up of:

- **Four ring-like "haem" groups**, each of which contains a central ion of iron (Fe^{2+}) – so a haemoglobin molecule contains four ions of iron, each of which can combine reversibly with an oxygen molecule.
- **A protein chain** called globulin.

Oxygen combines with haemoglobin to form bright red oxyhaemoglobin. This occurs in the blood capillaries of the lungs where there is a high amount of oxygen. Oxygen separates from haemoglobin at the cells of the body where there is a low concentration of oxygen (compared to capillary blood). At these sites, the oxygen diffuses into the cells of the body. Once inside the cells, the oxygen is used to produce energy. A byproduct of cellular energy production is carbon dioxide, and this is removed from cells to prevent its accumulation, which would cause the cells to become too acidic for cell enzymes to work, resulting in ill health. Carbon dioxide moves from the body cells into deoxygenated capillary blood, again by diffusion. Only about 15 per cent of the body's carbon dioxide is transported in the blood bound to the haemoglobin molecule (compared to 99 per cent

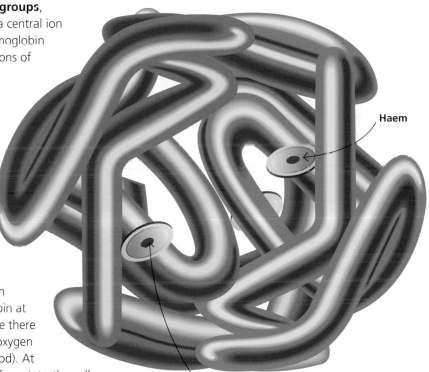

A computer model of a haemoglobin molecule.

Haem

Oxygen molecule binds to a haem group

of the oxygen). The carbon dioxide is bound to the amino acids of the global chain and not to the "haem" groups, thus colouring blood deep red.

Carbon dioxide is about 20 times more soluble than oxygen, so it dissolves easily in the watery plasma. Some 85 per cent of carbon dioxide is transported dissolved in plasma, and in red cells in the form of dissolved carbon dioxide gas, carbonic acid and bicarbonate ions.

Carbon Monoxide Poisoning

Carbon monoxide is a colourless, tasteless and odourless gas. If breathed into the lungs, carbon monoxide is highly toxic, since it binds very easily with haemoglobin. The red pigment binds much more efficiently to carbon monoxide than it does to oxygen. However, it is not the displacement of oxygen that is the main problem. The chief problem is that carbon monoxide poisoning causes haemoglobin to hold on to its oxygen, even in the capillaries at tissues, and so the tissue cells become deficient in oxygen, resulting in low energy production, which slows down cell activities, resulting in ill health.

Gases from car exhaust emissions contain high levels of carbon monoxide.

Asthma and Bronchitis

Asthma and bronchitis are disorders in which airway inflammation, excessive airway secretions and constriction of bronchioles increase airway resistance and reduces oxygenation of the blood.

Asthma is a respiratory disorder associated with erratic contraction of the bronchial smooth muscle (bronchospasm). This causes a shortness of breath, coughing and wheezing. The most frequent form of asthma is called extrinsic (atopic) asthma. This condition is a hypersensitive immune response to environmental agents, such as pollen or dust particles. Extrinsic asthma mainly occurs in children who are more susceptible to obstructive problems because of the structure of their airways, but it can also occur in adults. The initial bronchospasm in the airways usually resolves itself quickly, but there can be another episode of bronchospasm a few hours later. The first-phase bronchospasm occurs in response to histamine in the airways, whereas the second phase occurs in response to other chemicals, such as prostaglandins. Other inflammatory responses also occur, and these increase the airway resistance to air flow. Intrinsic asthma may be promoted by stress or exercise, and is found more typically in adults.

Bronchitis is an inflammatory condition of the airways, usually resulting from an infection or the presence of an irritant, such as cigarette smoke. Bronchospasm and fluid secretion are observed. With repeated incidences, excessive mucus production may be observed. A productive cough (one producing phlegm or mucus – sputum) lasting for more than 3 months in a year, and occurring 2 years in succession, indicates progression to chronic bronchitis. The chronic condition in particular has an overwhelming effect on the airway resistance, as smaller airways become blocked and mucus retention encourages infections. With time, a loss of structural components of the airway wall causes the collapse of alveolar tissue leading to the condition emphysema, which is severely disabling, due to the lack of oxygenation of blood and tissues.

Chapter 6
THE NERVOUS SYSTEM

The Communication Network

The communication network of the nervous system plays a key role in maintaining homeostasis, which is essential to your health and well-being. This network is constantly being challenged by the environment outside, but also inside the human body.

Your body must be able to detect both internal and external environmental changes, so that you can produce appropriate responses to achieve homeostasis. Your nervous system, together with the endocrine system, controls and coordinates all the body systems in order to maintain homeostasis.

The nervous system forms a communication network made up of all the neural tissue in the body. It is divided into two main parts – the central nervous system (CNS), which is made up of the brain and spinal cord, and the peripheral nervous system (PNS), which forms a network of nerves throughout the rest of the body.

The Peripheral Nervous system

The PNS carries sensory information to the brain and the spinal cord, and carries instructions back from the CNS to the rest of the body. The CNS is responsible for processing and interpreting the sensory impulses

Nerves spread throughout the entire body and provide a means of fast communication between different body parts.

Brain

Spinal cord

Radial nerve

Sciatic nerve

Femoral nerve

Tibial nerve

from the cells of the body, and coordinating the outgoing motor impulses to those cells to maintain homeostasis.

A large part of the information carried to the CNS comes from sensory receptors, which are found everywhere in the body, from complex organs, such as those of the cochlea of your ear, to "simple" cell membrane receptors on cells that are the "targets" for your hormones. In short, sensory receptors collect information about the external world, and also the environment within the human body around our cells. They convert this information into sensory nerve impulses, which are interpreted by the brain. These signals are carried by nerves, which form the widespread network of conducting pathways. The peripheral nerves are further divided into somatic and autonomic neurons. Somatic neurons carry signals to and from receptors all over the body. They are mainly concerned with your skeletal muscles, but also other voluntary muscles in the body, causing them to contract. They play an important role in coordinating an individual's posture and their movement, for example, maintaining your balance while instructing your skeletal muscle

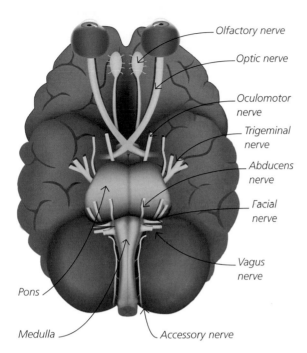

- Olfactory nerve
- Optic nerve
- Oculomotor nerve
- Trigeminal nerve
- Abducens nerve
- Facial nerve
- Vagus nerve
- Accessory nerve

Pons

Medulla

This image shows examples of the twelve pairs of nerves that enter the brain directly. They are called the cranial nerves.

during a running activity. Autonomic (visceral) neurons bring about automatic or involuntary changes. They coordinate the activities of most of the organs in the body, including blood vessels, and control your breathing and heart rates.

Peripheral Nerves

Peripheral nerves, which enter and exit the CNS by the spinal cord, are known as the spinal nerves. There are 31 pairs of these. The 12 pairs of peripheral nerves that enter the brain directly are called the cranial nerves. Each spinal nerve contains both incoming sensory and outgoing motor nerve fibres, from both the somatic and the autonomic nervous system. Spinal nerves enter and exit the spinal cord between the vertebrae and carry signals to and from specific areas of the brain. These nerves are not named individually, but according to their origin of the vertebra from which they emerge. For example, "lumbar 4" emerges from the fourth lumbar (lower back) vertebra.

Cranial Nerves

Cranial nerves, however, are named according to their activity. Your optic nerve, for example, carries signals from the eyes to the brain. These signals are concerned with the sense of sight, or vision. Cranial nerves may either contain incoming (sensory) neurons or outgoing (motor) neurons. Sensory cranial nerves are mainly from the "special" senses – the ears, eyes, nose and mouth, while cranial motor neurons mostly supply various facial muscles and the salivary glands. Some cranial nerves contain motor neurons, which are either somatic (voluntary) or autonomic (involuntary). For example, the voluntary control of the rectus muscles that move the eyes involves the abducens nerve (sixth cranial nerve), whereas the involuntary muscular movements associated with peristalsis in the throat involve the accessory nerve (11th cranial nerve). Other cranial nerves contain neurons that control both somatic and autonomic functions. The vagus nerve (tenth cranial nerve), for example, stimulates the voluntary movement associated with the swallowing process, and also the involuntary control of the heart rate.

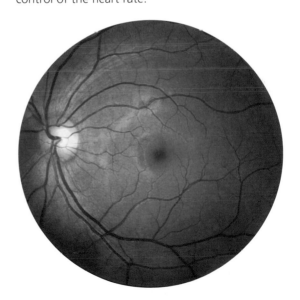

This photograph of the retina shows the point where the optic nerve leaves the eyeball (the yellow spot to the left) on its way to the brain.

Nerves and Nerve Cells

Nerves are channels for the collections of neurons, or rather their long processes. These processes are often simply called nerve fibres. In fact, the terms nerve cells, neurons and nerve fibres are often used interchangeably.

A nerve, therefore, might be viewed as nerve fibres bundled together rather like wires in an electrical cable.

There are three main types of neuron:

- **Sensory neurons** are so called because the information they carry originates at sensory receptors. These cells gather information from outside and inside the body and then convey signals to your CNS from all over the body.
- **Relay neurons** act as connectors between different neurons, such as sensory and motor neurons, passing signals between them.
- **Motor neurons** carry signals from the CNS to tissues elsewhere in the body to control voluntary and involuntary activities.

The structure of neurons in the CNS is similar to that of the PNS. All neurons contain cell organelles, such as a nucleus and mitochondria, but these cell parts are localized to a distended part called the cell body. The cell body has a number of fibre-like projections called dendrites, dendrons and axons, and branching terminals called synaptic terminals.

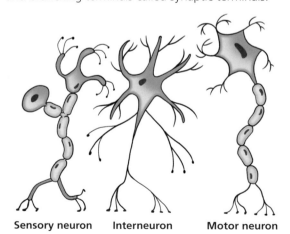

Sensory neuron   Interneuron   Motor neuron

Dendrites

Dendrites allow a fibre to communicate with other nerve cells that are in close proximity to it. Dendrons and Axons are longer projections of the fibre. They transmit electrical signals to and from the cell body respectively. Dendrites containing the synaptic terminals are where the neurons communicate with other nerve fibres, or muscle or gland cells in the distance, by releasing chemicals called neurotransmitters in the tiny gaps called synapses between these cell communities.

Secrets About Neurons and Glial Cells

- **The brain contains over 100 billion neurons.**

- **Some neurons in the central nervous system and in the retina of the eye have dendrites, but no axons.**

- **The longest sensory nerves are about a metre long and run from the toe to the bottom of the spinal cord. In contrast, many neurons in your brain and spinal cord are only 1 millionth of a centimetre long as their sole purpose is to send messages to neighbouring cells.**

- **Your brain contains between 10–50 times more neuroglia cells than neurons, and they make up to half the brain's weight!**

- **Myelinated neurons are like a super fibre optic broadband provider as they conduct the fastest nervous impulses.**

Loss of Nerve Supply to a Tissue

A single nerve may contain thousands of nerve fibres that supply more than one tissue to the brain and spinal cord. For example, the radial nerve (the name relates to the radius bone of the forearm) will contain nerve fibres that connect with the muscles of the forearm, the bone itself, joints of the limb, the blood and lymphatic vessels and the skin of the limb. Nerve trauma, therefore, will influence all of the tissues supplied by that nerve. Some regeneration of nerve connections is possible after trauma, but success depends upon whether the severed ends of nerve fibres are able to remake contact, and also, whether the surviving nerve fibres can branch to connect with cells that have lost their nerve connections.

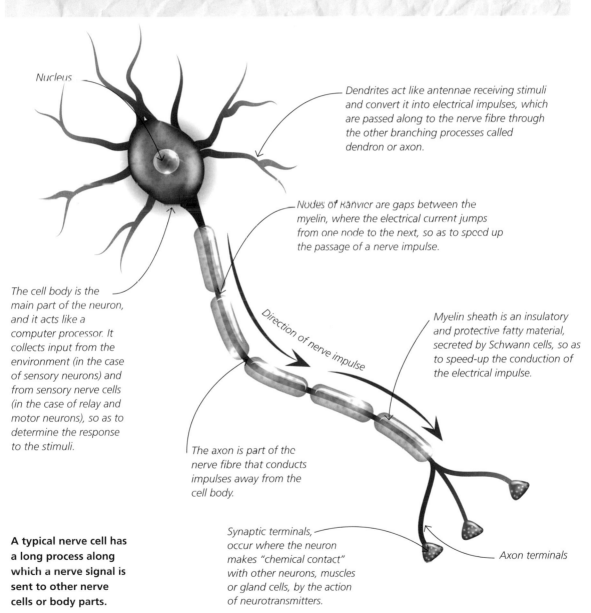

Nucleus

Dendrites act like antennae receiving stimuli and convert it into electrical impulses, which are passed along to the nerve fibre through the other branching processes called dendron or axon.

Nodes of Ranvier are gaps between the myelin, where the electrical current jumps from one node to the next, so as to speed up the passage of a nerve impulse.

The cell body is the main part of the neuron, and it acts like a computer processor. It collects input from the environment (in the case of sensory neurons) and from sensory nerve cells (in the case of relay and motor neurons), so as to determine the response to the stimuli.

Direction of nerve impulse

Myelin sheath is an insulatory and protective fatty material, secreted by Schwann cells, so as to speed-up the conduction of the electrical impulse.

The axon is part of the nerve fibre that conducts impulses away from the cell body.

A typical nerve cell has a long process along which a nerve signal is sent to other nerve cells or body parts.

Synaptic terminals, occur where the neuron makes "chemical contact" with other neurons, muscles or gland cells, by the action of neurotransmitters.

Axon terminals

How Nerve Fibres Communicate

Your nerve fibres are specialized to produce and dispatch electrochemical impulses. But how does this work?

Usually, receptors attached to sensory neurons pick up environmental cues or stimuli. Some receptors are designed to respond to changing cues in the body, such as changes in the heart and breathing rates, blood pressure and the fluid pressure of urine in your bladder. Others are designed to respond to changes in the environment outside the body, receiving external cues. These can include smell, sound, light, taste, pressure and temperature. However, not all stimuli are perceived – for instance, not all sounds are detected by the human ear. So the stimuli needs to reach a certain level (called threshold) before the receptors can convert the stimuli into an electrical impulse. This electrical current is known as depolarization and involves pumps in the cell membrane. These actively move positively-charged sodium and potassium ions in and out of the cell. The wave of depolarization passes along the nerve fibre until it reaches the synaptic terminals, where it triggers the release of the stored neurotransmitters.

Neurotransmitters

The neurotransmitters pass across the gap, called the synapse, between neurons to interact with receptors on the membrane of the next neuron

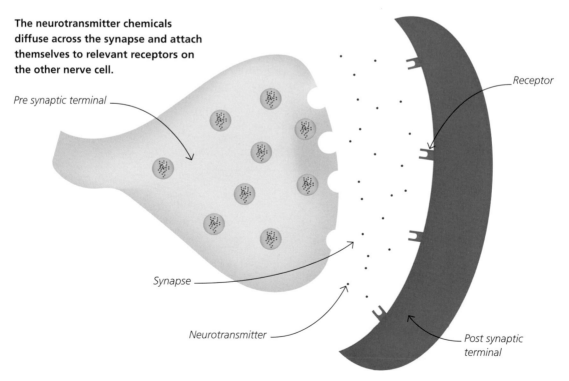

The neurotransmitter chemicals diffuse across the synapse and attach themselves to relevant receptors on the other nerve cell.

Pre synaptic terminal

Receptor

Synapse

Neurotransmitter

Post synaptic terminal

in the communication chain, known as the postsynaptic neuron. If enough stimulation occurs, an electrical impulse is produced, and a wave of depolarization passes along its nerve fibre.

Hopefully, you can see why an impulse is known as an electrochemical event and not just an electrical event, as it is called in lots of books! This is because if we cut an electrical wire to our TV, for example, then we will not get a picture, since the electricity current cannot pass to the TV. It is the same in the body. Electricity cannot pass over gaps that separate neurons, so synapses transmit information using chemical neurotransmitters. These chemicals are made in the nerve cell body and transported down the length of the axon to the synaptic terminals, where they are stored until they are needed to pass the impulse onwards.

The neurotransmitter is released into the synapse quickly and is either taken back to the neuron that produced it to be recycled, or the neurotransmitter is broken down in the synapse, so that the synapse is ready to respond to the next electrical signal it receives.

There is a neurotransmitter delay of about 0.5 milliseconds, which accounts for an electrical impulse arriving at a synaptic terminal and triggering off a response in the next neuron in the communication chain. This synaptic

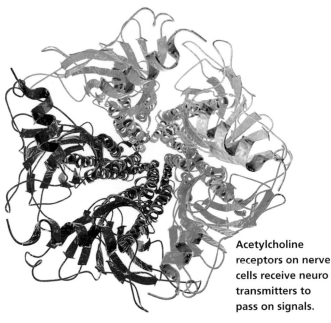

Acetylcholine receptors on nerve cells receive neuro transmitters to pass on signals.

neurotransmitter delay means the passage of the signal through a nerve pathway becomes much slower when more synapses are involved.

Nerve–Muscle Junction

A neuromuscular junction is a chemical synapse formed by the contact between a motor neuron and a muscle fibre. It is at the neuromuscular synapse that a motor neuron is able to transmit the signal to the muscle fibre, causing muscle contraction. Motor neurons also have this synapse with gland cells, such as salivary glands and sweat glands, across which the motor neuron signals the gland to release its secretion.

Secrets About Nerve Impulses

- **Your electrical Impulses pass down a nerve fibre at speeds of up to 125 metres (410 ft) per second.**

- **Your nerve can send up to about 1,000 impulses every second.**

- **Putting a cold pack on an injury reduces pain by slowing the speed of nerve transmission in nerve fibres associated with pain.**

- **The "gap" of a synapse is extremely narrow, and is just 20 nm (20 millionths of a millimetre) wide.**

- **A neuron makes 2,000–11,000 synaptic connections.**

- **The neurotransmitters are stored one side of the synapse to prevent information chaos, since the impulse can pass only in one direction.**

How is Your Central Nervous System Organized?

The neurons in the brain and spinal cord are similar to those of peripheral neurons. They have a cell body with dendrons and axons, and branching terminals. Likewise, synapses send signals to other cells.

Your brain contains some 100 billion neurons making the potential "circuitry" therefore, unimaginably extensive! Due to its structural complexity, the best way to illustrate the numerous activities of its component parts is through a series of illustrations, so this part of the chapter will focus on the external and internal views of the brain. The brain can be divided into four main regions: the cerebrum, the cerebellum, the diencephalon and the brainstem.

The Cerebrum

The cerebrum is the largest section of the brain and it is divided into four lobes – the frontal, parietal, occipital and temporal. Each lobe has distinctive areas associated with specific activities. The cerebrum is split down the middle into two halves called the cerebral hemispheres, and the two halves communicate through the corpus callosum.

The outer part of the of the hemispheres, the cortex, contains nerve cell bodies that are arranged

The parietal lobes are where we perceive touch and pain. These lobes also deal with information regarding the position of our limbs in space (proprioception) and where we interpret taste sensations.

The frontal lobes are where your personality is located. These lobes are involved in speech, emotions, thoughts, as well as skilled activities, judgement and social behaviour.

The occipital lobes are where you interpret and form visual images and where you recognize different colours. These lobes are also involved in hearing.

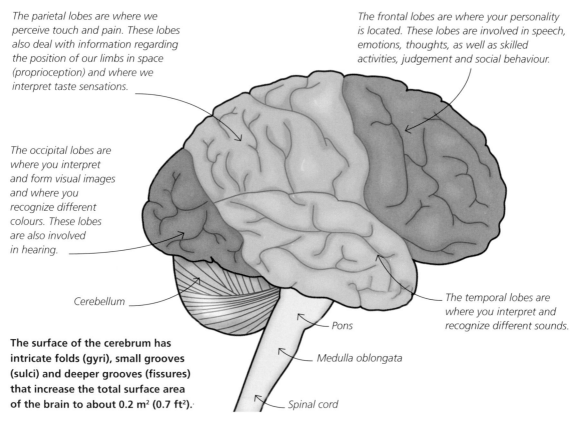

Cerebellum

The temporal lobes are where you interpret and recognize different sounds.

Pons

Medulla oblongata

The surface of the cerebrum has intricate folds (gyri), small grooves (sulci) and deeper grooves (fissures) that increase the total surface area of the brain to about 0.2 m² (0.7 ft²).

Spinal cord

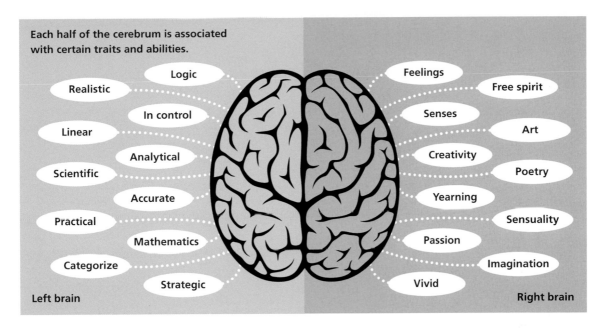

Each half of the cerebrum is associated with certain traits and abilities.

Left brain: Logic, Realistic, In control, Linear, Analytical, Scientific, Accurate, Practical, Mathematics, Categorize, Strategic

Right brain: Feelings, Free spirit, Senses, Art, Creativity, Poetry, Yearning, Sensuality, Passion, Imagination, Vivid

Left brain

Right brain

into layers. This is where the brain interprets sensations, initiates movements and carries out processes involved in thinking, speaking, writing, calculating, creating, planning and organizing.

The Cerebellum

The cerebellum is a large structure located behind and below the rest of the brain. In general structure, it resembles the cerebrum, hence it is often called the "little brain". It has a cortex of "grey matter" that connects with a set of subcortical structures called the cerebellar nuclei, known as the basal ganglia, which include the thalamus, putamen and caudate nuclei. These nuclei are involved in the control of complex movements, such as walking and running. The internal capsule of the cerebellum is a fan-shaped collection of myelinated axons (white matter) that connect the cerebral cortex to the brainstem and the spinal cord. It transports information to control movement in the upper and lower limbs. The cerebellum receives sensory information from the eyes, vestibular apparatus of the ears and from sensory receptors around the body. The cerebellum also communicates with the midbrain, providing an output that gives a finely-tuned modulation of activity that produces muscle contraction during

movements. The cerebellum, therefore, helps to coordinate movement, particularly the production of smooth movement, and fine movements such as those involved in writing and playing sport.

Down to the Stem

The diencephalon, the smallest part, is located beneath the cerebrum and on top of the brain stem in the centre of the brain. Within the diencephalon are two key brain structures: the thalamus and the hypothalamus. The thalamus acts as a relay station for nerve impulses coming from the senses, routing them to the appropriate part of the brain where they can be processed. The hypothalamus has a number of important functions, including the regulation of body temperature, hunger and thirst. The hypothalamus also acts as a link between the endocrine system and the nervous system, as well as controlling the release of hormones from the pituitary gland.

The brain stem sits on top of the spinal cord and is responsible for regulating the autonomic activities of the body that are vital to life, such as controlling the heart rate, the rate and depth of breathing, blood pressure and the sleeping and waking cycles. It is also important in the reflexes of swallowing and vomiting.

The Meningeal Membranes

The entire brain and spinal cord are covered and protected by three membranes, collectively called the meninges. The membranes extend downwards from the top surface of the brain to the opening at the base of the skull to reach the level of the second sacral vertebra.

The dura mater is a thick, tough protective and supportive layer made up mainly of the protein collagen. It contains two layers – an outer layer, attached to the skull, and an inner layer, known as the meningeal layer. The meningeal layer is continuous between the brain and spinal cord, and forms inward-folding membranes that project into the larger indentations of the brain, helping to support it. The most prominent of these indentations extends down and operates between the two hemispheres. Regions called venous

sinuses return blood flowing out of the brain to the veins in the neck, may be found between these two layers.

Middle Arachnoid Mater

This is a thin, delicate and transparent membrane that forms a loose layer, which helps cushion the brain and spinal cord. A narrow space separates this mater from the dura mater above, and this is known as the subdural space. In places, the arachnoid mater penetrates the dura mater and

The meninges surround and protect the brain, even down between the two halves of the cerebrum.

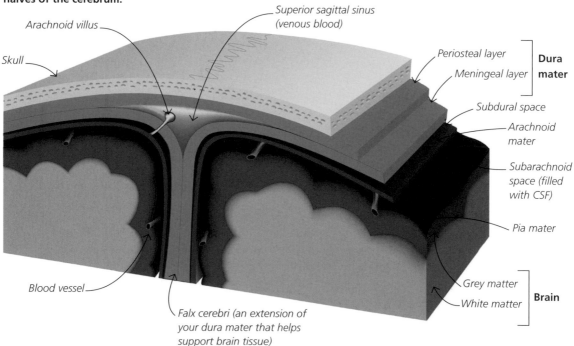

Arachnoid villus

Skull

Superior sagittal sinus (venous blood)

Periosteal layer — Dura mater
Meningeal layer

Subdural space

Arachnoid mater

Subarachnoid space (filled with CSF)

Pia mater

Grey matter — Brain
White matter

Blood vessel

Falx cerebri (an extension of your dura mater that helps support brain tissue)

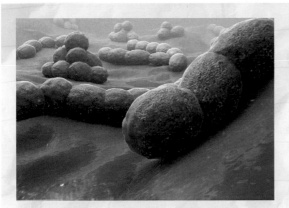

Streptococcus pneumoniae bacteria can cause meningitis.

Meningitis

This is an inflammation of the meningeal membranes caused by infectious agents, such as bacteria, viruses, fungi and parasites, or by the actions of microbial toxins. Pain receptors are found in the meninges, and severe headaches are a common feature of this infection. Irritation of cranial and spinal nerves could explain associated symptoms, such as neck stiffness, tinnitus and head retraction. In this inflammatory response, the meningeal vessels become more permeable, and if they secrete too much fluid, this will raise intracranial pressure as a consequence, which may also cause the above symptoms.

projects into the blood sinuses. These projections are the arachnoid villi (villus = finger-like projection) that reabsorb cerebrospinal fluid (CSF) back into the blood.

A larger space, the subarachnoid space, separates the arachnoid mater from the underlying pia mater, and contains the CSF that bathes the brain and spinal cord.

Inner Pia Mater

This delicate membrane covers the actual surface of the brain and spinal cord, and follows the contours of the cortex. The pia mater is rich in blood vessels that supply the neural tissue beneath. It also lines the cerebral ventricles (spaces or cavities), and forms the choroid plexus, the membrane that secretes the CSF.

Secrets About the Brain

- Estimates of the number of cells that each brain cell interacts with suggest that a single neuron may be directly associated with 10,000 to 11,000 others.

- Over 85 per cent of your nerve cell bodies are concentrated in the outer part of the hemispheres – this is your so-called grey matter.

- Curiously, the cortex of the right hemisphere receives most of its sensory information from the left side of the body, and the left hemisphere receives information from the right side. So if you are right-handed, you will have a dominant left hemisphere, while if you are left-handed then you will have a dominant right hemisphere. Some psychologists, however, see hemispheric domination as a myth.

- The lobes of the brain are named after the bones of the skull that sit over them.

- Your brain tissue may contain huge amounts of nervous tissue, but it cannot perceive pain.

1.5 kg (3.3 lbs)

The adult brain weighs about 1.5 kg (3.3 lbs).

Inside the Brain

Beneath the grey matter of your brain, which contains your nerve cell bodies, is the white matter. Internally, sections through the brain reveal this matter is made up of axons surrounded by glial cells that form myelinated sheaths, creating its white appearance.

Large parts of white matter link different parts of each cerebral hemisphere to each other, and to other parts of the brain. Inside the brain, there are also fluid-filled spaces called ventricles (not to be confused with the ventricles of the heart).

Cerebrospinal Fluid

The fluid is a special form of tissue-fluid called cerebrospinal fluid (CSF), as previously mentioned, and this circulates around the brain and the spinal cord. It helps to maintain a precise cellular environment (homeostasis) to optimize CNS activity.

The cerebrospinal fluid protects the brain against mechanical damage as well as providing it with nourishment. By their nature, neurons are particularly susceptible to changes in their environment, and even the normal fluctuations of blood glucose that arise through eating patterns can be disturbing for them as they require a constant glucose level. To ensure this, the central nervous system must, to a large extent, be physically isolated from the blood.

It is the CSF that provides the constant medium that bathes cells of the brain and the spinal cord. This fluid remains physically isolated from blood

The major structures of the inside of the brain.

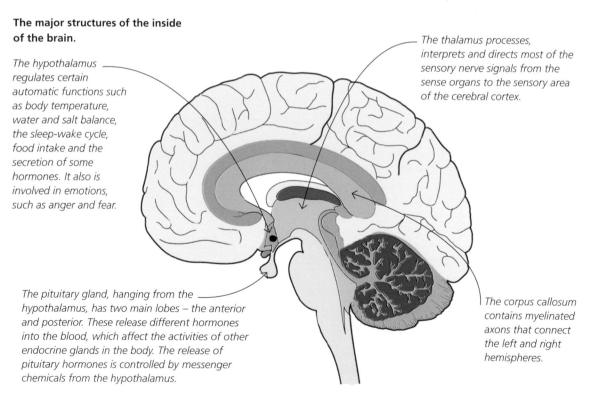

The hypothalamus regulates certain automatic functions such as body temperature, water and salt balance, the sleep-wake cycle, food intake and the secretion of some hormones. It also is involved in emotions, such as anger and fear.

The thalamus processes, interprets and directs most of the sensory nerve signals from the sense organs to the sensory area of the cerebral cortex.

The pituitary gland, hanging from the hypothalamus, has two main lobes – the anterior and posterior. These release different hormones into the blood, which affect the activities of other endocrine glands in the body. The release of pituitary hormones is controlled by messenger chemicals from the hypothalamus.

The corpus callosum contains myelinated axons that connect the left and right hemispheres.

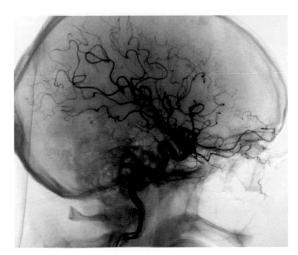

Blood vessels carry oxygen and nutrients deep into the brain and to the brain's cells.

Blood Supply to the Brain

A circle of arteries, called the "Circle of Willis", named after the English physician Thomas Willis, supplies the blood to the brain. This circle of vessels includes the anterior communicating artery, the left and right anterior cerebral arteries, the left and right internal carotid arteries, the left and right posterior cerebral arteries and the left and right posterior communicator arteries. From the circle of Willis, arteries branch to supply the brain cells. Tiny gaps between cells in the walls of the brain capillaries allow oxygen, glucose, water and dissolved substances to pass through to the brain cells, but prevent bacteria and some drugs getting through. This protective mechanism is called the blood-brain barrier.

Blood drains from the brain through small veins that empty into blood sinuses (spaces) within the dura mater. The largest of these is called the superior sagittal sinus. This blood, along with spent CSF, is discharged into the blood vessels in the neck before it returns to the heart.

because it is secreted by cells of the pia mater that line the larger fluid spaces. By doing this, the fluid composition can be very closely regulated, and this helps to protect the neurons from the moment-to-moment fluctuations observed in the composition of blood.

There are four main fluid ventricles – two lateral ventricles and the third and fourth ventricles. Most CSF is secreted into the lateral ventricles, and these connect with the third ventricle, which is connected to the fourth ventricle by the cerebral aqueduct inside the midbrain. The CSF flows from the ventricles into the subarachnoid space and then around the brain. Some passes into the subarachnoid space of the spinal cord and so circulates around the cord neurons.

Cerebrovascular Accidents

A cerebrovascular accident (CVA) or stroke occurs when a part of the brain is deprived of blood, and hence of oxygen and glucose. Cerebral thrombosis, cerebral embolism and cerebral haemorrhage are all common causes of CVA and may cause loss of neurons or an increase in fluid pressure on underlying nerve cells, which results in damaging areas of the brain.

Secrets of the Brain's Blood and CSF

- **Blood flow through the adult brain averages about 1000 ml (1.8 pints) per minute!**

- **The volume of CSF fluid at any one time circulating in an adult brain is about 120 to 150 ml (4 to 5 fl oz) and you produce about half a litre (almost a pint) of cerebral spinal fluid each day.**

- **A deficiency of CSF influences neural activities in the brain and headaches are a common symptom.**

- **Your CSF is normally crystal-clear and colourless; if it becomes viscous (sticky), it may indicate the presence of pathogens, as occurs in meningitis.**

The Brain's Secret Activities

The two halves of the cerebrum are connected by the corpus callosum. This allows communication between the two, and links similar positions of the left and right hemispheres and the rest of the body, through bundles of nerve fibres that cross from one side of the body to the other.

The right cerebral cortex receives sensory information from the left side of the body and sends motor information that controls movement on the left side. Likewise, the left side of the brain relates to the right side of the body.

Your sensory cortex contains sensory areas, motor areas and association areas. The sensory area receives and interprets information from the sense organs and other receptors throughout your body. Your motor area regulates your skeletal muscle movement, while the association area analyzes information received from the sensory areas, and fine-tunes these instructions before sending them to the motor area. Association areas are involved in thought and comprehension. They analyze experiences and interpret them in a logical way to make you fully conscious and aware.

Different parts of the cerebral cortex are responsible for controlling and interpreting different activities.

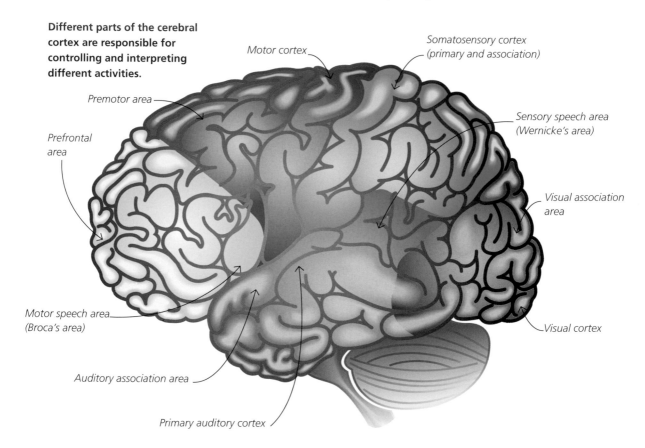

Motor cortex

Somatosensory cortex (primary and association)

Premotor area

Sensory speech area (Wernicke's area)

Prefrontal area

Visual association area

Motor speech area (Broca's area)

Visual cortex

Auditory association area

Primary auditory cortex

The Nervous System

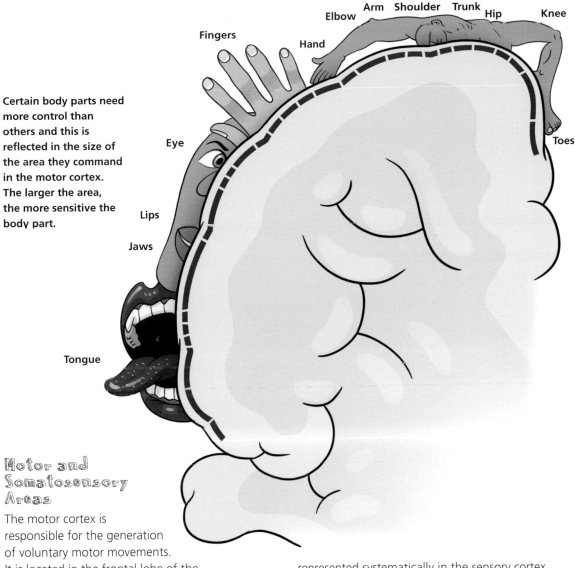

Certain body parts need more control than others and this is reflected in the size of the area they command in the motor cortex. The larger the area, the more sensitive the body part.

Fingers
Hand
Elbow **Arm** **Shoulder** **Trunk** **Hip** **Knee**
Eye
Toes
Lips
Jaws
Tongue

Motor and Somatosensory Areas

The motor cortex is responsible for the generation of voluntary motor movements. It is located in the frontal lobe of the brain, near a region known as the precentral gyrus. Both hemispheres of the brain have a motor cortex. Different parts of the motor cortex exert control over different parts of the body. These regions are arranged logically next to one another, for example, the region that controls the actions of the foot is next to the region that controls the leg and so on.

The somatosensory cortex sits along a region known as the postcentral gyrus. It receives sensory information from receptors in the skin including touch, pain, pressure and temperature from all areas of the body's surface. The body surface is represented systematically in the sensory cortex, with the head areas represented at the bottom of the postcentral gyrus, and the legs and feet at the top, so it is like a map of the body surface, only upside down!

The image above illustrates the relative proportion of the body reflecting the space these neurons occupy within the brain. For example, the hands and facial muscles are richly supplied with motor neurons to control their complex movements. The equivalent large areas in the somatosensory area, are associated with a large supply of sensory neurons to the lips, hands, feet and genitals, making them extremely sensitive.

Signals from colour photoreceptors are interpreted by the visual cortex and turned into a full-colour image.

have a variety of language skills, such as speech, reading, comprehension and writing. Therefore, a number of brain areas are involved in language. For example, the visual centre is involved in reading, the auditory centre in speech comprehension, and the motor centre in writing and speaking. Unlike the visual and auditory cortex, which are found in both hemispheres, language is restricted to the left side of the brain. These areas are known as Broca's area, involved in speech production, and Wernicke's area, which is involved in comprehension.

Visual and Auditory Areas

The visual cortex receives input directly from the eyes and the auditory areas from the ears. Damage to these areas can lead to blindness and deafness. These areas are called the primary visual cortex and primary auditory cortex. However, for full visual perception, additional processing in neighbouring cortical areas, called the secondary visual areas, is needed. It is in these areas that sensation is converted into perception. We know this because damage to the secondary visual areas does not lead to blindness, but can lead to loss of specific aspects of vision. An example of this is the condition achromatopsia, which is a loss of ability to see colour, so everything is viewed in black-and-white.

Other specific activities, such as language and memory, are located in precise areas of the brain. There is no single language centre, since humans

Other Brain Activities

Each of the billions of neurons within your brain makes neural connections with up to 20,000 other neurons, making the potential routes that your electrochemical impulses can take within your body and brain exceptionally large. It is these connections that control the higher functions of the brain, such as consciousness, intelligence and emotions, and make humans quite unique. The limbic system influences subconscious, innate behaviours related to survival, as well as our emotions. Many of these behaviours are reshaped by learned moral, social and cultural traditions. One part of the limbic system, the hippocampus, is involved with learning, recognition of new knowledge and memory. The limbic system is also linked to the sense of smell, which is why some odours stir up strong emotions and memories.

The limbic system consists of several key parts of the interior of the brain.

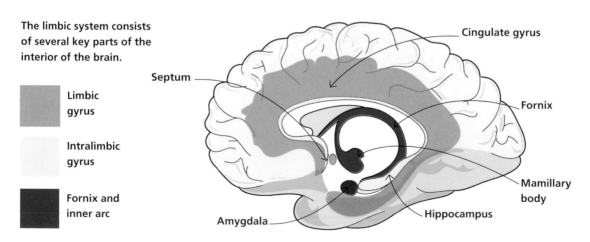

Limbic gyrus

Intralimbic gyrus

Fornix and inner arc

Cingulate gyrus

Septum

Fornix

Mamillary body

Amygdala

Hippocampus

The Nervous System

Memory, Intelligence and Emotions

The term "intelligence" means "understanding". It refers to the mental potential to reason, plan, solve problems, predict outcomes and think about complicated concepts such as time. It also involves the use of language and the skill to learn from past experiences. Intelligence is linked to the degree of folding or convolutions within the cerebral cortex within certain areas, such as the temporo-occipital lobe in an area called the posterior cingulate gyrus. So some believe that if you have more folds in those areas then you are more intelligent.

Memory is the ability to store, retain and, subsequently, remember information. It allows us to learn how to solve new problems by thinking about similar issues in which we have previously achieved a successful outcome. Memory involves many different areas of the brain, such as the hippocampus and mamillary bodies, which are related to different types of memory, including short-term memory, long-term memory, spatial memory and emotional memory.

Humans experience many different emotions, such as pleasure, anger, fear, happiness, sadness, curiosity, surprise, love, acceptance, compassion, anxiety and grief, to mention just a few. Emotions appear to result from the interaction of mind and body responses. Many different parts of the brain are, again, involved in the processing of emotions, especially the hypothalamus and parts of the limbic system.

More About the Brainstem

The brainstem initiates and coordinates many subconscious bodily processes that enable your body to function. It lies under the cerebrum and stems from the spinal cord. All nerve signals into and out of the brain from all parts of the body pass through the structure, and the nerves that connect the right and left sides of the brain cross over at this point. The main areas of activity in the brainstem are the midbrain, the medulla and the reticular formation.

Alzheimer's Disease

Alzheimer's disease produces a loss of reasoning, abstraction, language and memory, and the failure of such cognitive functions, and is frequently referred to as "dementia". The most obvious changes in brain structure are the formation of plaques and neurofibrillary tangles. The plaques are extracellular deposits of amyloid beta-protein, while the latter are dense tangles of protein fibres present within the cytoplasm of certain neurons. The plaques and tangles are not unique to the disease, however, as both occur to a lesser degree in the brains of elderly people. An understanding of the aetiology (causation) of Alzheimer's disease could, therefore, give information as to how plaques develop in the ageing brain.

Secrets About the Cerebral Cortex

- **Ninety per cent of people are right-handed. The remaining 10 per cent are left-handed or able to use both hands equally well (ambidextrous)!**

- **In a right-handed person, there are an estimated 180 million more neurons on the left side of the brain than on the right. In left-handed people this figure is reversed.**

- **Individuals with a damaged Broca's area typically speak very little or produce laboured speech, while an individual with damage to Wernicke's area will have poor speech comprehension and also often produce almost meaningless speech.**

- **Brainstem death is when the person no longer has any brainstem functions and has lost the potential for consciousness and the capacity to breathe.**

The Spinal Cord

The spinal cord is located in a hollow tube running through the vertebral column from the large opening in the skull (foramen magnum) to the second lumbar vertebra. It is covered by the three meningeal layers that also cover, protect and nourish the brain.

The whole cord is made up of 31 segments, each exposing a pair of spinal nerves; the segments and nerves are named after their corresponding vertebra. However, a group of spinal nerves from the second lumbar to the coccygeal nerves, hang loosely in the cerebral spinal fluid which surrounds the central nervous system. These are known as the *cauda equina*, since it resembles a "horse's tail".

The Body's Broadband

The spinal cord is a very complicated neural information superhighway. The main activities of the spinal cord are to provide a means of transmitting sensory activity from around the body to the brain, and motor activity from the brain to the tissues. Therefore, the nerve fibres of the spinal cord are said to be structurally and functionally connected through neural pathways to the brain. There are three types of these fibres:

- **Ascending pathways** that pass nerve fibres up the cord to the brain
- **Descending pathways** that pass nerve fibres down the cord from the brain
- **Interneurons** (or relay neurons) that connect ascending and descending neurons.

Many of the ascending and descending neurons pass along the periphery (outer part) of the cord and are myelinated, so they appear white in colour, and are known as the "white matter" of the spinal cord. The cell bodies of these nerve fibres, however, are located centrally, and the absence of myelin around the cell body structure makes this region appear darker – hence they are known as the "grey matter" of the spinal cord.

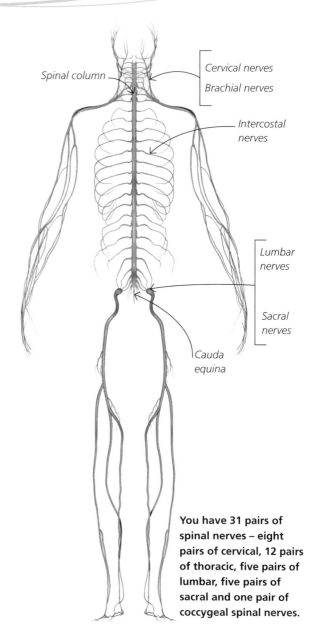

Spinal column

Cervical nerves

Brachial nerves

Intercostal nerves

Lumbar nerves

Sacral nerves

Cauda equina

You have 31 pairs of spinal nerves – eight pairs of cervical, 12 pairs of thoracic, five pairs of lumbar, five pairs of sacral and one pair of coccygeal spinal nerves.

The Nervous System

Spinal Reflexes

Spinal reflexes are responses that do not involve brain activity to start them. The main benefit of having spinal reflexes is that responses to sensory stimulation can occur much more quickly than they would if processing involved the brain. This is shown on the right by the withdrawal reflex of a limb in response to a pain. This requires contraction of muscles that, when stimulated, will move the limb away from the stimulus. For example, when you touch a hot object, this stimulates pain receptors in your fingers, which convert this stimulus into an electrochemical impulse along a sensory neuron to the spinal cord. Here, the sensory neuron synapses directly with the motor neuron (through interneurons) which, when activated, causes the biceps muscle to contract. This is a simple reflex arc. The neurotransmitter that is released in the synapse will be an excitatory one. Note that the withdrawal of the arm has been entirely processed by the spinal cord, although information regarding the pain stimulus will be transmitted to the brain, which responds by sending motor impulse to the voice box so you can vocalize your response to this painful stimulus by saying "ouch". This is called a secondary reflex and integrates spinal and brain activity. The role of the

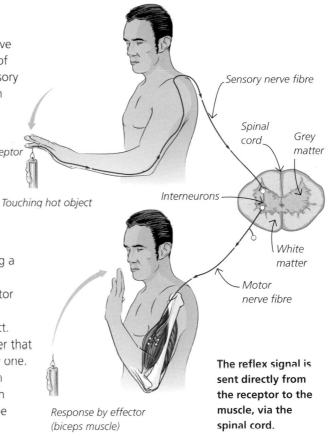

Receptor

Touching hot object

Sensory nerve fibre

Spinal cord

Grey matter

Interneurons

White matter

Motor nerve fibre

Response by effector (biceps muscle)

The reflex signal is sent directly from the receptor to the muscle, via the spinal cord.

simple withdrawal reflex is evident, since failure to withdraw your hand could potentially result in more damage to its tissues.

Secrets About the Spinal Cord

• **The term "grey matter" of spinal cord is inaccurate because the tissue only appears grey post-mortem – in life it is pink.**

• **The spinal cord is an extension of your brain and, in an adult, it measures about 45 cm (18 in) long and weighs about 35 g (1.2 oz). It contains about 1 billion neurons.**

• **The spinal cord stops growing in childhood, but the vertical column protecting it continues to grow.**

• **The nerves that emerge from the spinal cord to supply the arms and legs, merge to form complexes called nerve plexuses.**

• **Each spinal and cranial nerve subdivides into a number of branches that supply distinct parts of the body. The skin supplied by each nerve can be mapped out to form the dermatomes.**

The Special Senses

The special senses of touch, smell, sight, taste and hearing allow us to perceive and interact with the outside world. These "special" senses have their receptors organized into the sense organs – the eye, the ear, the nose, the tongue and the skin.

The special senses of vision, hearing, smell, taste and touch for the head region are located in the head. As such, their electrical activity does not pass through the spinal cord, but, instead, goes directly into brain tissue via the cranial nerves. The neural activity either passes directly to the thalamus, or travels to other parts of the brain, particularly inside the brainstem and hypothalamus, before passing to the cerebral cortex.

Hearing

Hearing, together with the sense of motion and balance, involves the stimulation of special receptors within your inner ear. "Sound" is the perception of small pressure waves generated in air (and water). The vibrations set up alternate compressions and decompressions of the air and are responsible for producing two different features of sound.

- **"Pitch"** (tone) is a measure of the number of compressions per second (or cycles per second) and is measured in hertz (Hz). This gives sound its frequency (or tones).
- **"Intensity"** (or loudness) is created by the amplitude of the compressions, and is measured using decibels. Rustling leaves, for example, have a decibel rating of 15 while a conversation is 45. The ear is extremely sensitive and can normally detect

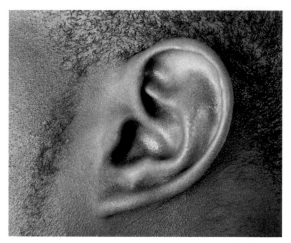

The skin and cartilage of the outer ear form a funnel to guide sound compressions into the inner ear.

intensities as low as 1 decibel and frequencies over a range of 20–20 000 Hz.

Hearing is, therefore, a complex means of transmitting pressure waves to the receptor cells. However, it is the pattern of distortion of these receptor cells that seems to be important in conveying information regarding sound intensity and frequency. Brain pathways that perceive sound are connected to the limbic system, so hearing sounds can involve powerful emotional responses, such as those felt by a parent when a baby cries.

The Process of Hearing Can Be Summarized As:

pressure waves → outer ear (tympanic membrane) → middle ear (ossicles) → inner ear (auditory receptor) → vestibular nerve (transmission) → neural pathways (processing cerebral auditory cortex)

The dark pupil is the hole through which rays of light enter the eyeball.

Seeing Things

Vision is one of your most important senses, allowing you to see and react to the environment around you. It involves detecting the intensity of light (brightness) and its wavelength (colour vision). About 70 per cent of your body's sensory receptors are in your eyes. As well as being able to see in bright sunlight, you can view faint starlight meaning a brightness range of 10 million to 1.

Photoreceptor cells have visual pigments that absorb light and break down into component chemicals that activate the cell membrane of the photoreceptor which, in turn, stimulate an electrochemical impulse in the optic nerve.

The Process of Vision Can Be Summarized As:

light → photoreceptor → optic nerve → neural pathways → interpretation
(retina)　　　　　(nerve　　　　(thalamus)　　　　(cerebral visual cortex)
　　　　　　　　transmission)

Smell and Taste

The senses of smell and taste begin with chemicals interacting with the smell (olfactory) and taste (gustatory) chemoreceptor cells. These send signals up to the brain. Unlike other chemoreceptors, such as carbon dioxide chemoreceptors, olfactory and gustatory cells respond to a vast array of chemicals. To be detected, these chemicals need to be dissolved in water provided by nasal and saliva secretions. The two senses are not as separate as people think. In fact, olfactory chemoreceptors are an important aspect of taste, even though taste is thought to be a feature of the mouth!

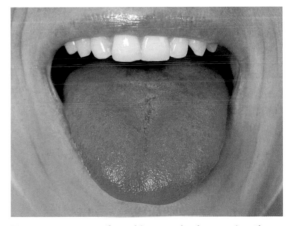

Taste receptors are found in taste buds covering the surface of the tongue and the back of the throat.

The Process of Taste Can Be Summarized As:

chemical → gustatory receptor → facial nerve transmission → interpretation
(tongue)　　　　　(brainstem/thalamus)　　　　(cerebral primary sensory cortex)

Pain receptors (called nociceptors) are located in the lining of the nasal cavities. These are stimulated by the irritating of odours, such as peppermint. Sneezing and tear-secretion are reflexes that may follow the stimulation of these fibres, and they are defensive mechanisms against a perceived harmful vapour. Adaptation to an odour occurs quickly, for example, a progressive build-up in gas from domestic appliances can cause needless fatalities.

With every breath, you bring in different aromas towards the smell (olfactory) receptors at the roof of your nasal lining. Unlike the receptors involving other senses, olfactory receptors are directly connected to your brain. Messages are passed to the limbic system, and therefore, as with taste, smells can also evoke powerful emotional responses.

The Process of Smell Can Be Summarized As:

chemical → olfactory receptor → facial nerve transmission → interpretation
(nasal cavities)   (brainstem/thalamus)   (cerebral primary sensory cortex)

Touch and Other Senses

The sensory receptors within the dermis pick up stimuli such as touch, cold, heat and pain. Touch receptors appear all over the body, but some areas of the skin, such as the palms and soles, have more receptors, so they are more sensitive to a tickle.

Other senses include equilibrioception (balance), proprioception (an awareness of the position of different parts of the body in space and relative to each other) and kinesthesia (joint movement).

Stroking a dog triggers touch signals to the brain.

The Process of Touch Can Be Summarized As:

chemical → Meissner's receptor → spinal/cranial nerves → interpretation
(dermis)   (spinal cord, brainstem/thalamus)   (cerebral primary sensory cortex)

The Nervous System

Chapter 7
THE ENDOCRINE SYSTEM

The Chemical Coordinators

The key activities of the endocrine and the nervous systems revolve around their ability to sense changes both inside and outside the body, and to bring about responses to these changes to maintain homeostasis.

The endocrine and nervous systems work in tandem. So much so, that they are often compared to orchestra conductors coordinating all the other body system activities. The neurons produce electrochemical messages, whereas the ductless glands of the endocrine system produce the chemical messengers called hormones.

Secret Commands

These messages coordinate activities of the body including your growth and development, repair and replacement of injured cells, control when you eat and when you stop eating, the production of sex cells (sperm and egg), the stress response, the sleep-wake cycle, fluid balance and even your mood swings! To coordinate these activities, your body needs to produce just the right amount of these messages on a second-to-second basis. The production rate of these messages, therefore, cannot be static, since they must be responsive according to the needs of the individual at any given time. Although some tissues are able to exert limited intrinsic control on some of their activities, the overall regulation of cells, tissues and the integration of organ functions are provided by these coordinating systems of the body.

Hormones control many of the body's activities, including the sleep cycle, rates of growth and how you respond to stress.

Like Aesop's tortoise and hare, the speed of nerve and hormone signals varies greatly, with super-speedy nerve signals and more sedate hormone signals.

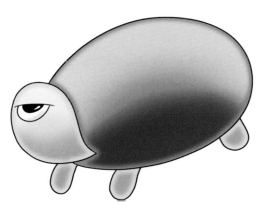

So let's consider the advantages and disadvantages of the each of these two coordinating systems in order to place hormones in the context of regulation of body activities (called homeostasis).

Comparing the Two Systems

Hormones are produced in specific endocrine cells in one part of the body, then released, or secreted, and transported in the blood to their specific hormone-receptor sites on "target" cells elsewhere in the body. On the other hand, nerve cells extend from one part of the body to another, so they provide a direct communication link. Hormone and chemical messengers (called neurotransmitters) from neurons combine with receptors at a cell level to coordinate their activities.

The distances travelled by neurotransmitters are tiny, and this is why nerve cells send "messages" (i.e., impulses) very quickly to the target cells. This means that cell activities can be changed within fractions of seconds. The disadvantage is that this direct link must be maintained, as neural damage can irreversibly prevent communication, and thus cell-directed activities. In contrast, the time is takes to produce, secrete and transport hormones to

their target cells, means that responses are much slower, and can take one or two hours, depending on the hormone. Hormones, therefore, are important in regulating medium- or long-term homeostatic activities of tissues, such as controlling body fluid composition or growth. Hormone operation would clearly be inappropriate, for example, in the rapid change necessary to control blood pressure at the moment when we stand to get out of bed or when we change from a resting to an activity-based exercise.

These two coordinating systems do not always act in isolation, and both systems are sometimes necessary in regulating the activities of the same organ. For example, rapid neural responses are needed in the fast coordination of gut movements, while the slower response to hormones is more suitable for controlling gut secretions (although neural activity can also alter these).

The overlap in actions of nerves and hormones is commonly illustrated by the stress response. The initial part of the response is called the "alarm" stage. Here the body's activities are altered firstly by activation of the sympathetic nervous system, which produces a rapid and immediate fright/fight/flight response. Following this initial burst of neural activity, the hormone adrenaline is released, to provide a back-up to the neural actions, so as to prolong the same physical activities associated with the alarm stage.

Hormones

The term "hormone" can be translated as "I excite", which reflects the role of hormones as chemical messengers that alter the activities of cells. However, the actual definition is actually much more involved.

A hormone is a chemical messenger that is produced and secreted from cells in one part of the body, such as the thyroid, which secretes the hormone thyroxine, thus speeding up energy production. The amount released into the blood in response to a specific stimulus varies according to the strength of the stimulus. For example, thyroxine is released in small amounts when we are at rest. However, when you are exercising or in times of stress when you need more energy, thyroxine is released in higher amounts. Hormones usually act on a part of the body that is some distance from the site of secretion. For example, insulin secreted from the pancreas targets cells of the liver, skeletal muscle and fat cells to increase their uptake of glucose from the blood, and to decrease the blood sugar level.

Types of Hormone

There are two main classes of hormones:

Peptide hormones, such as insulin, glucagon and oxytocin, are made from amino acids. These are soluble in water, but have poor lipid solubility. Do you remember from chapter 1 that the membranes of cells are mainly composed of lipids? Therefore, this means that peptide hormones bind to specific identity receptors on the outer surface of their target cell's membrane to produce their desired response.

Steroid hormones, such as testosterone, oestrogens, cortisol and aldosterone, are related to cholesterol, and, being modified lipids, they are highly lipid-soluble. So this means that these hormones pass through the cell membrane and bind to their specific identity receptors within the cell's cytoplasm to produce their desired response.

The hormone-receptor complex activates (switches "on") or inactivates (switches "off") specific genes associated with the hormonal activity. When genes are switched "on", they produce enzymes to stimulate a cell activity. In contrast, when a gene is switched "off", they do not produce enzymes, so an activity does not take place. This is how hormones coordinate body activities and control homeostasis.

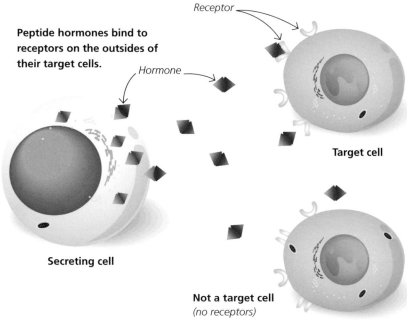

Peptide hormones bind to receptors on the outsides of their target cells.

Receptor

Hormone

Target cell

Secreting cell

Not a target cell
(no receptors)

Naming Hormones

Hormones are often named after their actions. For example, thyroid-stimulating hormone, is a hormone which stimulates the thyroid gland to release thyroxine. Some hormones are often referred to by a collective name, such as the tropins. These are released from the pituitary gland, which is sometimes called the "master" gland, since the tropins stimulate other endocrine glands to produce their hormones. For example, adrenocorticotropin is released into the blood from the pituitary gland and targets the adrenal gland cells to release its hormones, such as cortisol. The pituitary, however, would not be able to work without the gland called the hypothalamus.

So these two glands act as a single, coordinated functional unit, since the hypothalamus mediates the secretion of the hormones by the pituitary gland. Other endocrine glands include the pineal, thyroid, parathyroid, thymus, adrenal glands, pancreas and the female ovaries and male testes. The stomach and parts of the small intestine also produce hormones, which trigger the release of digestive enzymes from the pancreas, and the bile from the gallbladder. They also produce hormones that stimulate gut movements. The kidneys also have hormone-producing cells.

Hypothalamus produces hormones that stimulate or inhibit the secretions of the pituitary gland.

Pituitary gland, known as the "master" gland, secretes hormones that trigger the production of other endocrine organs of the body.

Pineal gland secretes a hormone that helps to regulate your sleep/wake cycles.

Thyroid gland secretes hormones that stimulate energy production within cells and lower blood calcium levels.

Thymus gland produces a hormone that stimulates the production and maturation of white blood cells called T lymphocytes.

Pancreas produces hormones that control blood glucose levels.

The main endocrine organs of the body (male and female)

Adrenal glands secrete many hormones, including sex hormones, stress hormones and hormones that control electrolytes in the blood.

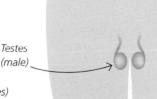

Testes (male)

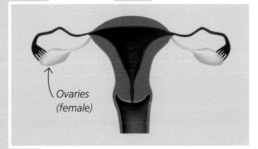

Ovaries (female)

Gonads (testes and ovaries) secrete hormones that control the development of the secondary sexual characteristics at puberty.

The Hypothalamus and the Pituitary Gland

Endocrine glands that send hormone signals to each other in sequence are referred to as an axis. Your hypothalamus and pituitary form an axis with a number of other endocrine glands.

The hypothalamus is part of the brain that links the endocrine system and the central nervous system. It has a diverse range of functions, including the control of "drives" of human nature (feeding, drinking, sexual behaviour), and the control of body temperature. The hypothalamus shares some activities with other parts of the brain, particularly with areas of the limbic system that are responsible for your emotions, anxiety and aggression. The hypothalamus lies at the base of the brain, directly below the thalamus, hence its name.

The pituitary gland is a small, pea-like projection from the hypothalamus at the base of the brain. It has two main lobes – the anterior pituitary and the posterior pituitary. It is located just below the hypothalamus, attached to it by a stalk – the infundibulum, sometimes called the pituitary stalk, a much easier name to pronounce.

Nerves and Hormones

Being a part of the brain, the hypothalamus contains nerve cells. Some of these are referred to as neuroendocrine cells, because they perform activities associated with both nerve cells and gland cells. The glandular cells secrete their hormones when stimulated, but being of neurological origin, the hormones may also be influenced by higher brain centres. As a result, many of the psychological influences on physical function, such as stress, are caused by the activities of the hypothalamus.

Most of the hormones produced by the hypothalamus are released into small blood vessels that form a direct link between the hypothalamus

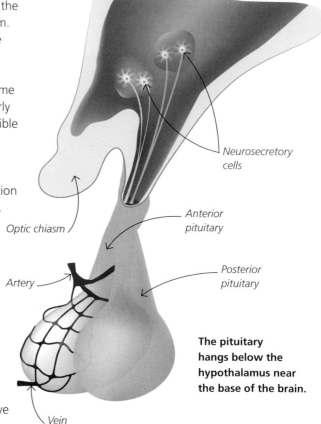

Neurosecretory cells

Optic chiasm

Anterior pituitary

Posterior pituitary

Artery

Vein

The pituitary hangs below the hypothalamus near the base of the brain.

and pituitary gland. These hormones are carried to the cells within the anterior lobe of the pituitary gland, so they can stimulate or stop the production of further hormones. So the hypothalamic hormones involved are referred to as either releasing or inhibiting hormones (or factors), according to their actions on the pituitary gland. They are usually named after the pituitary hormone that they influence. For example, thyroid-releasing hormone, secreted from the hypothalamus, controls the secretion of thyroid-stimulating hormone from the anterior pituitary.

The blood vessels that pass from the hypothalamus to the anterior pituitary gland are examples of portal vessels. This means that they carry blood directly from one capillary bed to another without passing through a vein or artery. Other hormones from the hypothalamus are secreted directly from nerve endings within the posterior lobe of the pituitary.

Most hormones secreted by the anterior lobe are released in response to the presence of hypothalamic-releasing hormones. However some, such as growth hormone, melanocyte-stimulating hormone and prolactin, have a dual regulatory mechanism. Their release is dependent on which hormonal mechanism dominates – the hypothalamic-inhibitory hormones, or the hypothalamic releasing hormones.

The posterior lobe consists of axon terminals that contain storage vesicles of two hormones – antidiuretic hormone and oxytocin. These are produced by the cells within the hypothalamus. Antidiuretic hormone, ADH (also called vasopressin) causes the collecting duct of the kidneys to reabsorb water and reduces the volume of urine, hence its name. Therefore, ADH is involved in the regulation of water balance within the body. It can also cause a rise in blood pressure, so the hormone is released in response to low body water (dehydration) and/or in response to low blood pressure.

Oxytocin is released during pregnancy; it causes the contraction of the uterus in the labour stage of childbirth, until the delivery of the baby. Oxytocin also has a role following birth, during lactation. It is responsible for milk secretion in response to the infant suckling the nipple.

The Secret of Negative Feedback

The hypothalamic-pituitary axis regulates the secretions from most of the other endocrine glands through a process called negative feedback. Using the hypothalamic-pituitary-thyroid axis as an example. When the blood levels of the thyroid hormone thyroxine are lower than normal, the hypothalamus releases thyroid-releasing hormone (TRH). This triggers the release of thyroid-stimulating hormone (TSH) from the pituitary gland, which in turn increases the output of the thyroxine. The production of TSH is also under negative feedback controlled by thyroxine, so TSH production falls as blood levels of thyroxine rise.

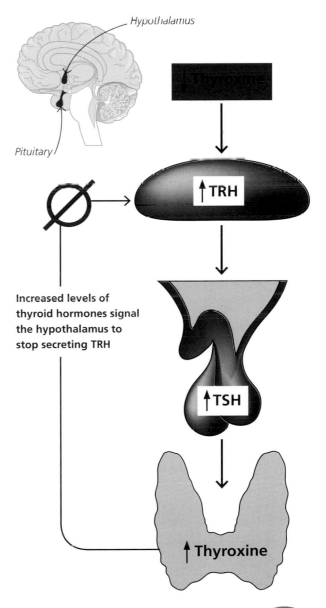

Increased levels of thyroid hormones signal the hypothalamus to stop secreting TRH

The Thyroid Gland

The thyroid gland is a butterfly-shaped gland in the neck. It has two main lobes that are connected in the middle. The thyroid is positioned just below the voice box and it overlaps the lower part of the windpipe.

The thyroid has a rich blood supply and its cells produce and secrete two hormones – thyroxine and calcitonin. Thyroxine controls the speed of chemical reactions within cells by stimulating energy production for these reactions. Therefore, this hormone is important for the growth and functioning of every system in the body. The main activity of calcitonin is to promote the uptake of calcium by bone cells when blood calcium is too high. As a result, it returns blood calcium to its normal range through a negative feedback mechanism. As blood calcium declines, the release

The thyroid gland sits in the neck, in front of the trachea, or windpipe. It holds a quarter of your body's iodine stores.

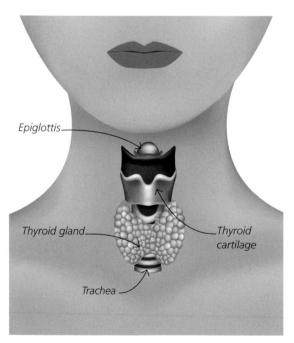

Epiglottis

Thyroid gland

Thyroid cartilage

Trachea

Secrets About the Thyroid and Parathyroid Glands

- **Thyroxine, also known as T4, is a weak hormone that is converted into a more active hormone called triiodothyronine (T3).**

- **Some people have only three parathyroid glands, while others have six.**

- **Parathormone promotes the activation of vitamin D within the kidneys.**

- **Vitamin D is now recognized by many scientific authorities to be a hormone (although it has retained its name), and it increases calcium absorption from food contents in the gut.**

of calcitonin is reduced. Calcitonin activity forms only one part of the process by which blood calcium is regulated. Its levels reflect a balance between uptake from the intestine, uptake or release from bone tissue, and removal in urine.

There are usually four parathyroid glands, on the back surface of the thyroid gland. These glands secrete a hormone, called parathormone. This raises your blood calcium level when it falls below its normal levels. It does this by increasing the release of calcium from your bone stores, increasing calcium reabsorption from the filtered fluids in your kidneys, and by increasing calcium absorption from your gut.

The Endocrine System

The Pancreas

The pancreas is often called the "mixed gland", since it has both endocrine and exocrine cells. Its exocrine (or digestive) cells secrete pancreatic juice into the duodenum of the small intestine.

The role of pancreatic juice is covered in the digestion chapter. The endocrine cells release hormones directly into the blood and pancreatic hormones are concerned with maintaining blood glucose levels. The main endocrine cells (called the Islets of Langerhans) are the beta cells, which produce the hormone insulin and alpha cells, which, in turn, produce the hormone glucagon.

Insulin is released into blood when there is too much glucose in the blood (hyperglycaemia). The insulin is transported to its "target" cells - skeletal muscle, liver and fat cells - where its actions are mediated through binding to insulin receptors on the membrane of these cells. This binding increases the transportation of glucose into these cells, thereby decreasing glucose levels in the blood. Conversely, when there a low level of glucose in the blood (hypoglycemia), glucagon is secreted into the blood and transported to glucagon receptors, which are on the same target cells as insulin. The glucagon-receptor binding results in glucose release from these cells, back into the blood, thereby raising its level in the blood. While insulin is the only known hypoglycemic hormone, glucagon is one of many hormones in the body that stimulates glucose release from cells.

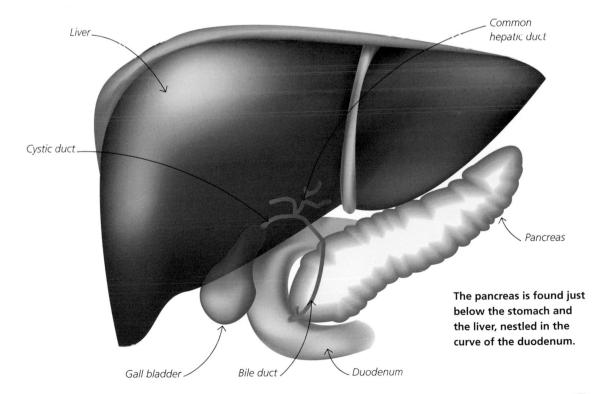

Liver

Common hepatic duct

Cystic duct

Pancreas

Gall bladder

Bile duct

Duodenum

The pancreas is found just below the stomach and the liver, nestled in the curve of the duodenum.

Diabetes Mellitus

Diabetes mellitus is a problem associated with insulin. This leads to hyperglycemia, which involves the removal of glucose in the urine (glycosuria) and dehydration. It can also cause urinary tract infections. When it goes on for a long time, hyperglycemia has the potential to damage the eye (retinopathy), the nervous system (neuropathy) and the renal system (nephropathy).

Diabetes mellitus can arise if the individual is incapable of producing insulin. This is called insulin-dependent diabetes mellitus (IDDM). It also occurs in an individual who has a poor tissue response to insulin, which is called non-insulin-dependent diabetes mellitus (NIDDM). The former condition typically occurs in children or adolescents, and the latter during adulthood. In fact, IDDM is sometimes still referred to by the earlier names, Type 1, early-onset or juvenile diabetes, while NIDDM may be referred to as Type 2, or late-onset, diabetes.

Care for people with IDDM revolves around the need to take insulin as a hormone replacement, and to pay particular attention to dietary habits and foods eaten, especially in relation to the timing of the individual's insulin injection. The aim is to provide extrinsic control of blood glucose concentration, thus preventing excessive hyperglycaemia or hypoglycaemia. Care for people with NIDDM has similar aims, and will involve dietary advice. However, this form of diabetes may also be treated with drugs, depending upon the actual problem. For example, a "stimulant" drug may be used to speed up the utilization of glucose.

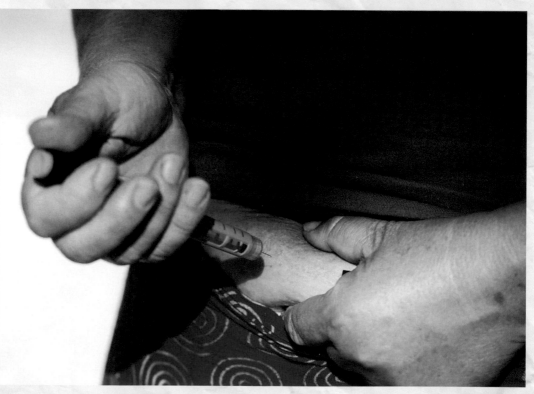

People suffering from diabetes may need to take regular injections of insulin to control blood sugar levels.

The Adrenal Glands

As the name suggests, the adrenal glands lie adjacent to the kidneys; in fact they lie on top of them. Each gland has an outer layer, or cortex, and an inner layer, or medulla.

The cortex secretes a range of steroids that are known as corticoids. These are the glucocorticoids, mineralocorticoids and gonadocorticoids.

The most important glucocorticoid is cortisol, which has a variety of roles in the body. The net effect of this hormone is to maintain glycogen stores, so it can be mobilized as and when required, and increase blood glucose and free fatty acid concentrations in blood, so as to increase their availability to various tissues when needed. This hormone is particularly useful during physical activity and following trauma, when wounds may need healing and to re-establish a "normal" state of homeostasis generally. The hormone is released when the body is under stress, which is why it is often called the "hormone of stress" and our ability to cope with stress is reduced by its absence.

The main mineralocorticoid is aldosterone, and its release is stimulated by sodium deficiency or by an increased plasma potassium concentration. It has an important role in the maintenance of sodium (and hence water) and potassium balance. Aldosterone targets cells of: the kidney, to stimulate sodium uptake and increase the excretion of potassium and hydrogen ions; the sweat and salivary glands, to help with sodium reabsorption from sweat and saliva; and the gut, to facilitate sodium absorption.

The adrenal glands are about 3 cm (1 in) wide, 5 cm (2 in) long and 1 cm (½ in) thick.

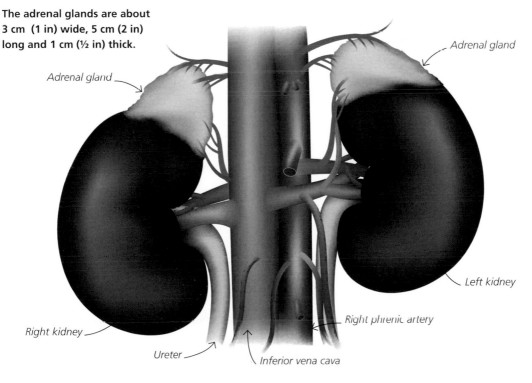

Adrenal gland

Adrenal gland

Left kidney

Right phrenic artery

Right kidney

Ureter

Inferior vena cava

Adrenaline and noradrenaline are responsible for the fright/fight/flight response to stressful situations.

Gonadocorticoids include the male androgens and female oestrogens. However, the production of these sex hormones by the adrenal gland is minor compared with that of the sex glands or gonads, and their actions for much of adulthood are probably of little consequence. Nevertheless, they do have important influences on the development of the unborn child, prepubertal growth during childhood and in the development of the secondary sexual characteristics of the male and female.

The adrenal medulla produces the catecholamine hormones adrenaline and noradrenaline. These are also known as stress hormones, because they are released in times of danger, when the sympathetic nervous system initiates the fright/fight/flight response. Their actions support and prolong the rapid changes designed to improve your chances of survival.

Secrets of the Pancreas, Kidneys and Adrenal Glands

- **Your pancreas is both an endocrine gland, making hormones such as insulin, and an exocrine gland, secreting digestive juices into the duodenum by the pancreatic duct.**

- **The consumption of alcohol decreases the production of antidiuretic hormone, causing you to urinate more and, as a result, you may become dehydrated and "hung over".**

- **Adrenaline and noradrenaline act both as hormones and as nerve cell communication chemical messengers called neurotransmitters.**

- **Blood levels of adrenaline and noradrenaline increase as much as a thousandfold within one minute during acute stress.**

The Gonads

The gonads (testes and ovaries) produce mainly steroid hormones. The male steroids are collectively called androgens, while the female steroids are oestrogens and progesterone.

The main male steroid hormone is testosterone, which controls the production of spermatozoa, while the female steroid hormones regulate the menstrual cycle and breast development. The release of male and female hormones is controlled by gonadotropins, called follicle-stimulating hormone (FSH) and luteinizing hormone (LH), from the anterior pituitary. These hormones are named after their effects in the female, but they are found in both sexes. They are released in response to the presence of gonadotropin-releasing hormones (GnRH) produced by the hypothalamus. Regulation is exerted by negative feedback effects of the gonadal steroids on the hypothalamus and pituitary.

The male gonads, the testes hang outside the body, where they are kept cooler to aid sperm production, while the female ovaries are found on either side of the womb.

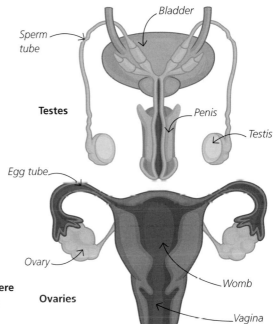

Sperm tube · Bladder · Testes · Penis · Testis

Egg tube · Ovary · Womb · Ovaries · Vagina

Menopause

During the menopause, there is a decline in the release of oestrogens, which causes a degeneration (atrophy) of the female reproductive organs, and other effects, such as "hot flushes" and sweats. The withdrawal of ovarian steroids may enhance the actions of male testosterone, which is produced in small amounts throughout life by the adrenal cortex of both sexes. This hormone may cause the growth and coarsening of facial hair and a deepening of the voice in women following menopause. However, in the long term, the most serious post-menopause effect can be the deposition of cholesterol in blood vessels, which increases the risk associated with heart disease and strokes, while the loss of bone material increases the risk of osteoporosis. Studies indicate that hormone replacement therapy (HRT) around and after the menopausal years prevents, or at least reduces, the risk of these changes.

It is debatable whether there is a male equivalent. However, although testosterone release and its breakdown decline with age, so plasma concentrations only decline slowly, if at all. Sperm production continues well into late adulthood, although the numbers of sperm and their ability to survive may be slightly decreased.

Endocrine Disorders

Endocrine disorders quite simply arise due to an inadequate secretion (hyposecretion) of a hormone or an over-secretion (hypersecretion) of a hormone. Specific examples are provided here, along with the general principles.

Hyposecretion can be caused by the endocrine cells lacking receptors, as in the underactive thyroid disorder called Hashimoto's disease. It can also be caused by an inability of the endocrine gland to produce the hormone, as occurs in insulin-dependent diabetes mellitus, or because of a deficiency. Alternatively, it may be caused by a deficiency in a dietary component which is needed for the production of the hormone, for example a lack of iodine in the diet may cause thyroxine deficiency.

Hyposecretion can sometimes be improved with drugs that stimulate production and/or secretion of the hormone from the gland. Often, however, correction requires hormone replacement therapy, and the replacement hormone is normally a synthetic form of the natural chemical.

In the future, it may be possible to insert endocrine stem cells, which can produce the hormone that is deficient or absent in hyposecretory disorders. The introduction of pancreatic Beta Islets of Langerhans is likely to be the first success with stem cell transplantation, as there is plenty of government funding into developing this therapy to treat insulin-dependent diabetes mellitus.

Hypersecretion frequently occurs when there is a failure of the negative feedback mechanism that controls hormone release. This can arise because there is a lack of receptors to the feedback signal to the gland cells or, because an overgrowth (hypertrophy) of the endocrine gland itself leads to overactivity of its cells.

Principles of correction might involve using a drug to suppress hormone production or to antagonize its actions. Availability of such drugs is limited at the present time. Surgery, involving partial or total gland removal to reduce the hypersecretion, is another option.

Robert Wadlow (1918–1940) was the tallest person in recorded history. By the time he died at the age of 22, he was 2.72 metres (8.9 ft) tall and his incredible size was due to high levels of human growth hormone caused by hyperplasia of his pituitary gland.

The Endocrine System

Chapter 8
THE DIGESTIVE SYSTEM

The Body's Food Processor

Your digestive system breaks down the food you eat into something that your body can absorb and use in order to survive.

Your digestive system is essentially a long tube that begins at one opening (your mouth) and ends at another (your anus). Its activity in food breakdown (or digestion) and food utilization is helped by organ chemicals (mainly enzymes), which speed up this breakdown. These organs are the pancreas, the gallbladder and the liver.

The digestive system is likened to a food-processor. It receives large insoluble chemicals at one end and then breaks them down, using a combination of mechanical and chemical means, into simpler, soluble nutrients. These nutrients are then absorbed into your blood and transported to your body cells. Here, nutrients are used in various chemical processes to maintain the health and well-being (homeostasis) of the body. For example,

The bulk of your diet is provided by three main types of food – carbohydrates, lipids and proteins. However, a good healthy diet should also include other smaller and soluble substances, such as water, vitamins, minerals and nucleic acids.

some nutrients are used to produce energy and for growth and repair of injured body parts. In short, absorbed nutrients ensure that cells receive the chemicals necessary to maintain their life, hence the old adage "you are what you eat"! However, this saying is not entirely true, since some components of food, for example fibre, are not digestible, and they are usually disposed of as tidy "parcels" at the anal end.

Is It All About Digesting Food?

The conversion of food into nutrients involves five main activities:

- **Ingestion** (eating) is the process of taking food into the mouth.
- **Digestion** is the mechanical and chemical breakdown of food into simpler nutrients.
- **Absorption** is the passage of these nutrients into the transporting (cardiovascular and lymphatic) systems, which carry nutrients to the cells that need them.
- **Assimilation** is the liver's role in maintaining the normal levels of these nutrients in blood, so that cells can take them in when needed to maintain their optimal activities.
- **Defecation** is the removal from the body of what remains from the food we consume, waste products of bile and unabsorbed substances, such as some water and electrolytes.

The physical breakdown of food involves the chewing action of the teeth and the squeezing action of gut muscles on food. The purpose of physical digestion is to break up the food into smaller parts, so as to increase the surface area to help with the process of chemical digestion.

The Digestive System

From the mouth to the anus, the digestive system of an adult human can be nearly 9 metres (30 ft) long.

The chemical breakdown of food is actually performed by water in the process called hydrolysis – digestive enzymes ONLY speed up this process. Water disintegrates the chemical bonds of complex foods, such as carbohydrates, proteins and fats, into simple soluble chemicals called sugars, amino acids and fatty acids respectively, so that these can be absorbed into the blood. Most nutrients are absorbed in the final part of the small intestine, called the ileum. Also, water is absorbed in the colon of the large intestine, leaving behind a solid waste for its removal from the body.

Physical and chemical digestion occur simultaneously. For convenience, however, the processes will be described individually for each part of the gut. But you should remember that it is the integration of these gut activities which determines the outcome of the digestive and assimilative processes.

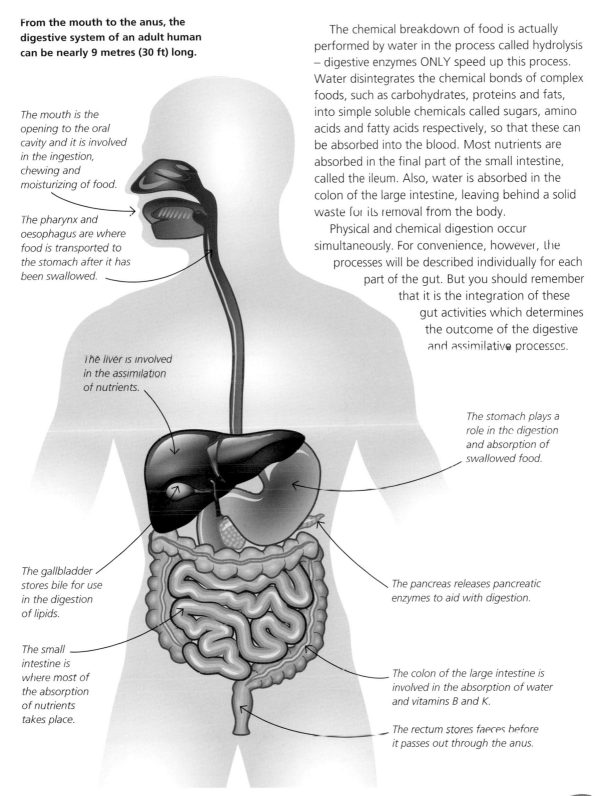

The mouth is the opening to the oral cavity and it is involved in the ingestion, chewing and moisturizing of food.

The pharynx and oesophagus are where food is transported to the stomach after it has been swallowed.

The liver is involved in the assimilation of nutrients.

The stomach plays a role in the digestion and absorption of swallowed food.

The gallbladder stores bile for use in the digestion of lipids.

The pancreas releases pancreatic enzymes to aid with digestion.

The small intestine is where most of the absorption of nutrients takes place.

The colon of the large intestine is involved in the absorption of water and vitamins B and K.

The rectum stores faeces before it passes out through the anus.

The Mouth

**The mouth is where the process of digestions begins.
Its structure allows for the ingestion, biting, chewing and
moisturizing of the food we consume.**

The teeth are involved in the first process of digestion – biting and chewing.

Adult teeth include the chisel-like incisors, the pointed canines and the flattened premolars and molars. Regardless of the type and different shapes, the structure of each tooth is the same. Each tooth has a crown, neck and root. The crown is the visible region above the gum. It is covered by the hardest substance in the human body, enamel, which stops it wearing down. The neck is the middle part of the tooth below the gum, which leads to the root. The main bulk of the tooth is a bone-like yellow substance, called dentine. This is the second-hardest substance in the body and it protects the teeth from breaking when chewing. The next internal layer is the pulp, located in the pulp cavity. This cavity contains blood vessels and nerves and, as such, it provides nutrients and sensations to the teeth. The root is located in a bony socket and is held in place by a ligament, fixed in the cementum (a soft version of bone that covers the dentine) and gums.

The incisors (biting teeth), along with the mouth opening, control the size of the food particles we take in. The canines historically were used by primitive humans for tearing fleshy meat from bone. However, since our diet and social eating habits have changed, these teeth have become less important and are smaller in modern humans. Once inside the mouth, the premolars and molars crush and grind food, breaking it up into smaller parts. This is controlled by the jaw muscles, and the mechanical process is called mastication (chewing). At the same time, food is mixed and moistened with saliva from the salivary glands.

The orientation of different types of teeth in an adult jaw.

Incisor

Molar

Premolar

Canine

Changing Teeth

The development of teeth begins before you are born, but they do not show until a few months after birth. These are your "milk teeth", and there are 20 of them. The incisors erupt first, then the canines and finally the molars. Your milk teeth are shed from about 7 years of age and replaced with "adult teeth", which are in place when you reach 13. A further four "wisdom teeth" may appear during teenage or early adulthood years. Excluding the wisdom teeth, there are 28 adult teeth, and these are made up of eight incisors, four canines, eight premolars and eight molars.

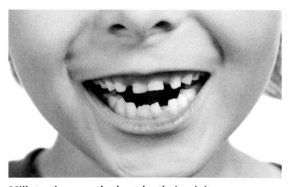

Milk teeth are pushed out by their adult successors.

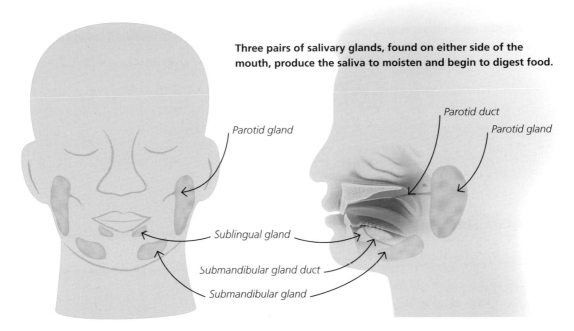

Three pairs of salivary glands, found on either side of the mouth, produce the saliva to moisten and begin to digest food.

Parotid gland

Parotid duct

Parotid gland

Sublingual gland

Submandibular gland duct

Submandibular gland

Chemical Digestion in the Mouth

In theory, salivary amylase is capable of helping to convert large, insoluble carbohydrates (starch) into the simpler sugars (maltose). However, in practice, the food is only in the presence of amylase during chewing and on its brief journey to the stomach. Food is usually in the stomach within 4 to 6 seconds of eating it. After that, the food is then within an acidic environment, and this inactivates the amylase (which works best in pH 7–8), causing carbohydrate digestion to stop. No other foods are chemically broken down in the mouth. As a result of the physical and chemical processes, the food leaving the mouth is reduced to a soft, flexible ball (called a bolus) that is swallowed easily.

Secrets About Enzymes and Saliva

- Did you know that without the actions of digestive enzymes, we would never receive the nutrients in a form that could be absorbed into blood?

- Did you also know that changes in acidity and alkalinity in different parts of the gut are vital, since the different enzymes need different pH levels to work most effectively?

- The largest salivary glands, the parotids, produce about 25 per cent of your daily secretion, which is between 1–1.5 litres (1¾–2½ pints), whereas the submandibular glands produce around 70 per cent, and the smallest, the sublingual, produce only about 5 per cent of your daily production of saliva.

- The mumps virus (myxovirus) infects the parotid glands, causing their enlargement and inflammation.

- The submandibular and sublingual glands are responsible for the spray of saliva that sometimes flows out when you yawn.

- Saliva glands cease to secrete saliva in states of dehydration, in order to conserve body water. That is why you have a dry mouth when you are dehydrated.

Swallowing and Peristalsis

The swallowing reflex involves voluntary and involuntary stages where food is guided down the back of the throat and into oesophagus, without entering the windpipe and causing choking.

The voluntary stage of swallowing occurs when the tongue voluntarily moves the bolus of food to the back of the mouth and then into the throat. The involuntary throat stage involves the swallowing reflex. Nervous impulses cause the soft palate to move upwards, sealing off the nasal passageways and preventing food from entering the nasal cavity. Nervous impulses also cause the larynx to move upwards, sealing off the opening of the larynx, called the glottis, with the epiglottis.

Occasionally, when we drink liquids very quickly, the sealing of the nasal passageways is too slow and the drink passes into and out of the nose. Alternatively, food particles may be swallowed so fast that the sealing of the glottis is incomplete and food becomes lodged at the top of the larynx. This stimulates the coughing reflex, which is usually enough to expel the bolus from the larynx and prevent you from choking.

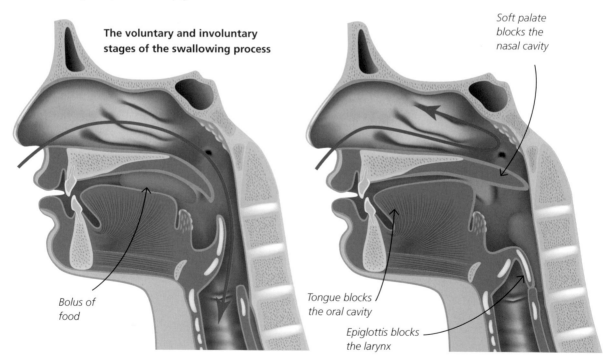

The voluntary and involuntary stages of the swallowing process

Soft palate blocks the nasal cavity

Bolus of food

Tongue blocks the oral cavity

Epiglottis blocks the larynx

While the airway remains open, the tongue voluntarily pushes the bolus of food towards the back of the mouth.

As the bolus enters the throat, the epiglottis involuntarily shuts over the larynx, blocking entry to the windpipe.

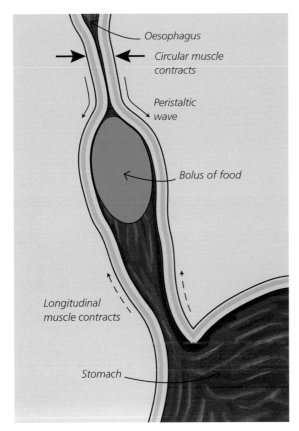
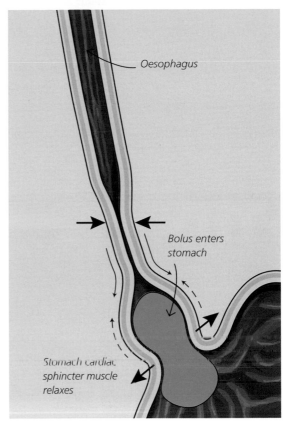

Smooth muscle tissue in the walls of the oesophagus contract to create waves of peristalsis to push the bolus down into the stomach.

The oesophageal stage begins as soon as the bolus of food has entered the oesophagus and waves of muscular movement, called peristalsis, are responsible for its transport down to the stomach. Peristalsis involves the simultaneous contraction of the circular muscle behind the bolus, which squeezes the oesophagus and forces the bolus downwards, and of the longitudinal muscle in front of the bolus, which shortens and expands the diameter of the section involved, and thus allows the forward movement of the bolus.

Swallowing also promotes the relaxation of the normally contracted sphincter muscle, which guards the entrance to the stomach and allows passage of the bolus into the stomach. Peristaltic movement of food occurs throughout the gut, starting in the oesophagus and ending in the anal canal, where it helps with the removal of faeces.

Dysphagia and Oesophagitis

• Dysphagia is a difficulty in swallowing. It can result from a mechanical obstruction of the oesophagus, for example caused by a tumour, or a disorder that impairs oesophageal motility, such as the occurrence of it in Parkinson's disease, or muscular disorders that interfere with voluntary swallowing or peristalsis.

• Oesophagitis is reflux of the stomach contents into the oesophagus. This is mainly due to inefficiency of the muscle sphincter that guards the entrance to the stomach. The acid that is present naturally in the stomach irritates the cell lining of the oesophagus, causing "heartburn".

The Stomach

The stomach is a J-shaped muscular organ, located immediately below the diaphragm on the left side.

The stomach's entrance and exit are guarded by rings of muscle called the cardiac and pyloric sphincters, respectively. These are normally contracted, so they stop any contents from leaking out. The cardiac sphincter opens when a bolus of food is in the lower part of the oesophagus, allowing it to enter the stomach. Simultaneously, the pyloric sphincter closes so that food cannot pass into the small intestine without first undergoing digestion in the stomach.

The size and shape of the stomach varies according to its content. The stomach's folds, called rugae, are clearly visible when the stomach is empty and they disappear when the stomach is filled with food.

Physical Digestion

Stomach churning is a three-dimensional muscular movement that is caused by overlapping layers of muscle that lie at right angles to each other, which is unique to this organ. It increases the efficiency of digestion by physically breaking down a food bolus into smaller pieces and by mixing the food with the stomach's chemicals (gastric juice), making chemical digestion more efficient.

Chemical Digestion

The main activity of stomach secretion, gastric juice, is to convert the semi-solid bolus of food from the mouth into a semi-liquid called chyme and start the breakdown of proteins. The lining of

Sitting just below the diaphragm, the stomach is positioned between the liver and the pancreas. It is about 30 cm (12 in) long and 15 cm (6 in) wide, at its widest point.

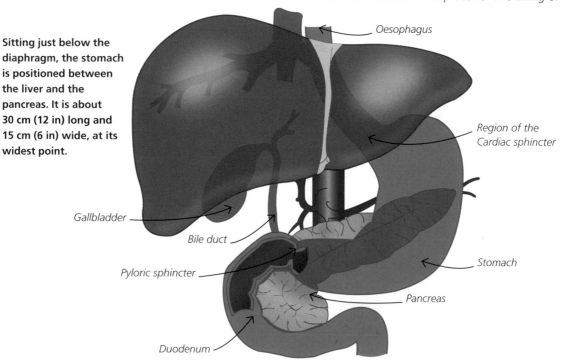

Oesophagus

Region of the Cardiac sphincter

Gallbladder

Bile duct

Pyloric sphincter

Stomach

Pancreas

Duodenum

the stomach contains gastric pits.

Parietal, or oxyntic, cells secrete hydrochloric acid and intrinsic factor, which is vital for the absorption of vitamin B12, which important for the production of red blood cells.

Chief, or peptic, cells secrete an inactive form of the enzyme pepsin (called pepsinogen), which requires hydrochloric acid for its activation. Pepsin helps water with the breakdown of proteins into chains of amino acids called polypeptides. Peptic cells also secrete an enzyme called rennin, which curdles milk by converting the soluble protein found in milk into an insoluble form of protein, so that the pepsin and water can break down this protein into polypeptides. This enzyme is abundant in babies and infants who are being breastfed.

Mucus cells produce mucin, which mixes with water to produce mucus. This sticks to the stomach's inner linings to prevent autodigestion of the stomach wall by the hydrochloric acid and proteolytic enzymes present in gastric juice.

How Long in the Stomach?

Food can stay in the stomach for between 2 and 4 hours, depending on its content. For example, a meal containing steak will have a large amount of protein and so will remain in the stomach for about 4 hours, whereas a bowl of chips which contains

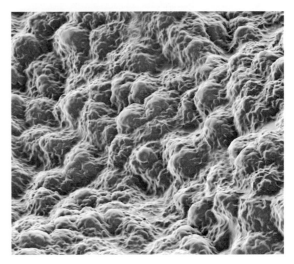

Gastric pits lining the stomach wall contain secretory cells that release a variety of substances that make up the gastric juice.

no protein, will spend less time in the stomach. After a couple of hours, the bolus of food is converted into a semi-digested creamy slurry called chyme. Once digestion is well on its way, regular wave-like contractions start to push the stomach contents down towards its exit at the pyloric sphincter, and then out into the duodenum. As more and more chyme passes into the duodenum, the stomach gradually gets smaller.

The adult stomach has a very small volume when empty, but can expand to hold 2 to 4 litres (3½ to 7 pints) of food when full.

Pyloric Stenosis

Pyloric stenosis is a disorder of gastric emptying. The pyloric sphincter grows excessively, causing a narrowing of the stomach's exit. As a result, an extra peristaltic effort is needed to force the gastric contents through the narrowed pyloric sphincter. The stomach's muscle layers may also become bulkier. Pyloric stenosis is more common in small babies and it usually presents itself about the third week following birth. The hallmark symptom of this condition is projectile vomiting – the spraying of liquid vomit some distance from the infant.

Secrets About the Stomach

• You produce about 2 to 3 litres (3½ to 5 pints) of gastric juice every day.

• Your stomach (as with the rest of the gastrointestinal tract) is composed of four main layers. Starting from the inside, the innermost (lining) layer is called the mucosa, and gastric juices are made in gastric pits here. The submucosa is the next layer, which is surrounded by the muscularis, a layer of muscle that moves and mixes the stomach contents. The final, outermost layer, the serosa, forms a "wrapping" of connective tissue for the rest of the stomach layers.

• The destructive hydrochloric acid in the stomach helps to kill alkali-liking microbes that are present in the food or that have been swallowed during a respiratory tract infection.

• To protect itself from this corrosive acid, the stomach lining creates a coating of mucus.

Vomiting

Vomiting is a forceful regurgitation, known as anti-peristalsis, of the contents of the gastrointestinal tract. It is a neural reflex involving the vomiting centre of the medulla of the brainstem. The reflex causes the epiglottis to seal the glottis and close the larynx, so the vomit does not pass into the airways. The nasal passageways are also closed to prevent the entrance of the vomitus into the nose, and the pyloric sphincter shuts, which increases stomach pressure, causing gastric regurgitation. These closures are accompanied by powerful contractions of the diaphragm and the abdominal wall muscles, pushing the vomit up and out from the gut.

Vomiting may occur in response to one or more of the following: intense fear, anxiety, unpleasant smells, gastrointestinal irritation by chemicals or microbes, brain tumour, certain drugs, such as morphine or general anaesthetics, or conflicting impulses from the organs of balance in the ear, as occurs in travel sickness.

Vomiting may involve removing potentially harmful substances from the body. For instance, vomiting 1 to 8 hours after a meal could indicate food poisoning.

The Small Intestine

Your small intestine extends from the stomach to the large intestine. It is a highly coiled structure occupying a large part of the abdominal cavity.

The small intestine is divided into three parts:

- **The duodenum**, which forms a loop below the stomach, encloses the pancreas. It receives bile from the gallbladder and pancreatic juice from the pancreas.
- **The jejunum** extends from the duodenum to the final part of the small intestine (the ileum). The duodenum and jejunum are both concerned mainly with digestion.
- **The ileum** connects the small intestine with the large intestine. It is the main area of absorption of nutrients.

Physical Digestion

The principal movement within the small intestine is called segmentation, which, as its names suggests, results in the breaking-up of food particles into smaller and smaller "segments". It also provides a thorough mixing of the food, now called chyme, with digestive juices from this part of the gut. Peristaltic movement is weaker than it is in the oesophagus, so food is kept in the small intestine for longer, reflecting the time required for digestion to be completed.

Bile, a yellow-green alkaline fluid, contains bile salts, and these convert large globules of fat into small droplets in a process called emulsification. The process is rather like pouring cooking oil into water. As the large globule of oil first enters the water, it disperses into smaller fat droplets, increasing the total surface area. This larger surface area aids the chemical digestive process, breaking fats down into parts that the body can absorb.

Your small intestine is about 6.5 m (21 ft) long and it is coiled inside the large intestine.

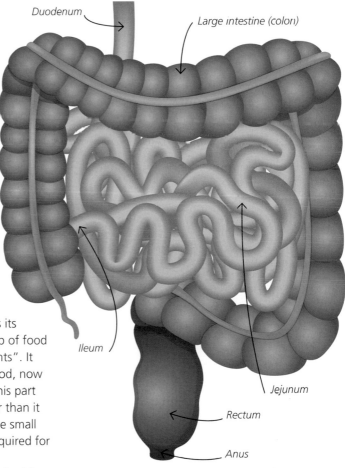

Duodenum

Large intestine (colon)

Ileum

Jejunum

Rectum

Anus

Chemical Digestion

The components of chyme must be digested chemically. The small intestine starts and completes these processes using the secretions of the pancreas and the intestinal lining, known as the mucosa. Most digestion occurs in the duodenum and jejunum.

Gallstones

The main cause of gallstones is the crystallization of excess cholesterol. Gallstones often go undetected in the body, but as they increase in size, they may be responsible for small, infrequent or even total obstruction to the flow of bile from the gallbladder and into the duodenum.

A partial obstruction to the outlet from the gallbladder results in a heartburn pain or discomfort. This is known as biliary colic, and it occurs when the gallbladder contracts after eating a meal. The resulting inflammation is called cholecystitis, reflecting the old name of the gallbladder, the cholecystic gland. If the stone becomes mobile and lodges itself, there is intense pain and fever, with the yellow coloration of the skin and eyes characteristic of obstructive jaundice. Complete obstruction of the flow may even be fatal.

Large gallstones can measure 5 cm (2 in) across.

The enzymes helping water with the chemical digestion of nutrients appear in brackets after each of the chyme components:

Carbohydrates

Starches (pancreatic amylase), sugars such as maltose or malt sugar (intestinal maltase), lactose or milk sugar (intestinal lactase) and sucrose, also called cane or table sugar (intestinal sucrase), are chemically reduced to the simple sugars that can be absorbed into the bloodstream. These simple sugars are glucose, fructose or fruit sugar, and galactose or grape sugar. They may also be present in food themselves, of course, and since they are small enough, they are directly absorbed into the blood.

Fats or Lipids

These have not been chemically digested up to this point. They enter the small intestine in their consumed chemical form (pancreatic lipase and intestinal lipase are the enzymes responsible their digestion). The chemical breakdown products of fats are fatty acids and glycerol. These are now absorbed into the blood and lymphatic circulation.

Proteins

Polypeptides (pancreatic trypsin) are present as a result of the protein digestion of gastric pepsin. Peptides (intestinal peptidases) result from the breakdown of polypeptides. At this point, protein digestion is complete, and the end-products of protein digestion – the nutrients called amino acids – can now be absorbed into the circulation.

Other Essentials

Vitamins, minerals and water, like simple sugars, are not digested because they are small enough to be absorbed across the gut wall.

The regulation of the physical and chemical digestive activities of the gut is a combination of neural and hormonal mechanisms sometimes controlled by the thought or the sight of food, but mostly by the presence of it in the gut.

Absorption

Absorption is the process whereby the end-products of digestion (nutrients) and other soluble nutrients, which do not need to be broken down, are transported from the inside of the gut into the body's transport systems.

The ileum is the main site of absorption. It has:

- **A large surface area**, as the ileum is long at around 6.5 m (21 ft), and the surface area of its lining is increased with circular folds (plicae) and finger-like projections, villi and microvilli.
- **A very thin lining** to aid absorption.
- **An extensive blood and lymphatic supply** to the villi.
- **A small extracellular space** between the absorptive cell and the blood capillaries and lymphatic vessels, so the nutrients only have a short distance to travel from the gut to the blood and lymph.
- **The walls of the blood** and lymphatic vessels consist of a very thin lining (endothelium), which allows rapid absorption.

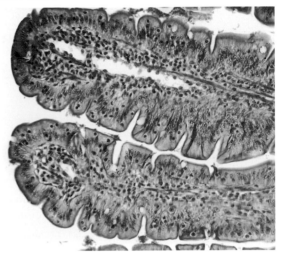

A micrograph showing a close-up of villi that line the wall of the small intestine.

Secrets about the Small Intestine

- The duodenum is the shortest section of the small intestine, extending about 25 cm (10 in), the jejunum is approximately 2.7 m (8.9 ft) long, and the ileum is around 3.6 m (11.8 ft) long. The diameter is 2.5 cm (1 in). In comparison, your large intestine is about 1.5 to 2 m (4.9 to 6.6 ft) long, but it has a much wider diameter than the small intestine.

- The pancreas produces and releases about 1200 to 1500 ml (2 to 2½ pints) pancreatic juice daily. Intestinal juice is produced at a rate of 2 to 3 litres (3½ to 5 pints) per day.

- Your bile is produced in the liver, about 80 to 100 ml (2.8 to 3.5 fl oz) daily, from where it travels to the gallbladder to be stored and concentrated, before it enters the duodenum.

- The alkaline pH of the intestine comes from the combined pHs of pancreatic juices, bile and intestinal secretions. It neutralizes the effects of the gastric acid as it enters the duodenum, but also provides an environment that works best for the enzymes.

- The ileum accounts for 90 per cent of absorption. The mouth, stomach and large intestine absorb the remaining end-products of digestion.

The Large Intestine

The large intestine extends from the small intestine to the anus. Its longitudinal muscle layer forms three bands, called taeniae, that are kept in a state of contraction, giving the large intestine its "pouched" appearance.

The large intestine has four main areas – the caecum, colon, rectum and the anal canal.

The activities of the large intestine include the storage of faeces until it is removed from the body, the secretion of mucus to ensure lubrication of the faeces and ease its removal from the body, and the absorption of most of the remaining water, mineral salts and the vitamins B and K.

Bowel Movements

Peristalsis within the colon is coordinated by the slow wave of the colon, which gets faster as it passes along the colon – like seeing a wave far out at sea which, when it gets closer to shore, becomes larger and more powerful. Out of the bowel contents received each day, only about 10 per cent, 200 ml (7 fl oz) of semi-solid waste remains for removal from the body.

The caecum is the first pouch of the large intestine which receives indigestible remains from the small intestine. Below it is the appendix.

The appendix is a blind-ended tube of lymphatic tissue and may play a role in gut immunity.

The anal canal is richly supplied with arteries and veins. Its opening to the exterior is called the anus, which is regulated by two sphincter muscles – the internal and external sphincters.

Faecal material enters the rectum for removal through the anal canal. When the rectum is distended with faeces, reflex contractions stimulate a powerful urge to open the bowels. These contractions are the internal anal sphincter involuntarily relaxing, forcing faeces towards the external sphincter muscle at the end of the gut. This sphincter is under voluntary control (following infant toilet training) and stays tightly shut, until you decide when the sphincter is to be opened.

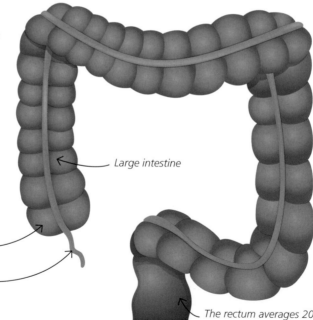

Large intestine

The large intestine runs around the outside of the abdominal cavity, encircling the small intestine.

The rectum averages 20 cm (8 in) in length, and it expands for the temporary storage of faecal material. Its final 3 cm (1 in) makes up the final part of the large intestine – the anal canal.

Anus

Abdominal quadrants

The abdomen can be divided anatomically into four quadrants. Each quadrant contains a number of digestive and other organs. Abdominal pain is often referred to by doctors by the quadrant within which it is localized. This knowledge may be used in conjunction with clinical tests to diagnose a disease in an organ in that quadrant.

The four abdominal quadrants

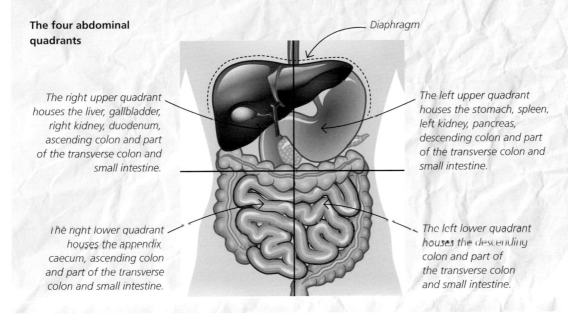

Diaphragm

The right upper quadrant houses the liver, gallbladder, right kidney, duodenum, ascending colon and part of the transverse colon and small intestine.

The left upper quadrant houses the stomach, spleen, left kidney, pancreas, descending colon and part of the transverse colon and small intestine.

The right lower quadrant houses the appendix, caecum, ascending colon and part of the transverse colon and small intestine.

The left lower quadrant houses the descending colon and part of the transverse colon and small intestine.

Peritoneum

The intestines as well as the abdominal and pelvic cavities are surrounded by a single transparent double-membrane sac called the peritoneum. The outer (parietal) peritoneum lines the walls of the cavities, while the inner (visceral) peritoneum covers the organs within the abdominal and pelvic cavities. A small amount of fluid separates the two layers. The peritoneum holds the abdominal contents in place and allows them to slide over each other without friction.

Both membranes on the back wall of the abdomen fuse to form double sheets, each called a mesentery. These mesenteries provide access for the blood and lymphatic vessels and nerves to supply the digestive tract, and also to stabilize the relative positions of each part of the digestive tract. This prevents the intestines becoming entangled during the digestive movements associated with peristalsis, churning or segmentation, or even during a sudden change in body position, for example when rolling onto your side in bed. Each mesentery contains a rich blood and lymphatic supply that helps to prevent infection entering the body through the intestinal tract. The mesenteries also provide a storage site for fat within the abdomen.

Peritonitis

Sluggish colonic movement is conducive to bacterial growth. These bacteria are potentially pathogenic if released into other areas of the body, such as the abdominal cavity, which can happen if the intestines are perforated. They may cause life-threatening acute inflammation of the peritoneum called peritonitis. A less serious form of peritonitis can result from the rubbing together of inflamed peritoneal membranes.

The Liver

Most of the liver is found in the right upper quadrant of the abdominal cavity under the protective cover of the lower ribs. Sitting on the left side of the abdomen, the liver lies just above the stomach and below the diaphragm.

The liver is soft, and extremely red due to its rich blood supply, which is why lacerations of the liver are dangerous, as the individual bleeds profusely and such injuries are difficult to repair. The liver is covered almost entirely by peritoneum. It has two main lobes, a large right lobe and a smaller left lobe, separated by ligaments.

Liver Lobules

The absorbed nutrients are transported into the liver inside the hepatic portal vein. This vessel subdivides into smaller vessels in the liver, and these enter a connecting system of smaller blood spaces or channels called sinusoids. The liver's functional units are called the liver lobules. A hexagonal capsule surrounds each lobule. Internally, each lobule consists of chains of liver cells, known as hepatocytes. The chains are arranged radially around a central vein. Between the chains are the blood and bile sinusoids, which allow an exchange of substances between hepatocytes, blood and the bile channels. The blood sinusoids carry oxygen-rich hepatic arterial and nutrient-rich hepatic portal blood from the edge of each corner of the hexagonal capsule to the central vein. This vein is a tributary of the hepatic vein and it drains blood from the liver into the inferior vena cava and then back to the heart.

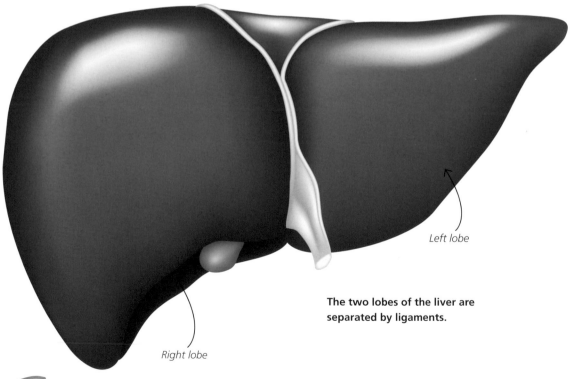

Left lobe

The two lobes of the liver are separated by ligaments.

Right lobe

Both lobes of the liver are served by a rich blood supply from the hepatic artery and the Portal vein.

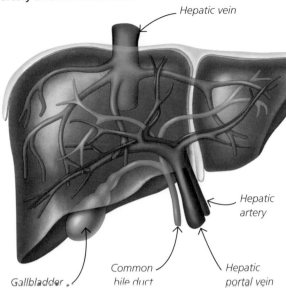

Hepatic vein

Hepatic artery

Hepatic portal vein

Common bile duct

Gallbladder

The bile channels carry bile produced from the hepatocytes lining them. This bile drains along a series of channels (ducts) from the liver into the gallbladder. Here, it is stored and concentrated until it is needed by the small intestine for the process of emulsification of large fats.

Assimilating Nutrients

One of the main jobs of liver cells is to assimilate absorbed nutrients. These include:

- **Glucose**, which combines with oxygen for cellular respiration. If absorbed glucose causes a high blood sugar, then any excess glucose is stored as glycogen. This can then act as an emergency fuel supply, releasing glucose when the blood sugar levels are low, as may occur during exercising and overnight fasting.
- **Fatty acids** and glycerol, which are used to make cholesterol and triglycerides.
- **Amino acids**, which are used to make transport proteins, such as albumin, which transports lipids and blood clotting proteins, such as fibrinogen. If there are excess amino acids they are ultimately converted to urea and glucose. Urea is a waste which is removed from the body in urine and sweat. Excess fatty acids can also be processed to make glucose, if the diet is deficient in this carbohydrate. Otherwise any excess fat is stored under the skin and around vital organs.
- **Fat soluble vitamins** – A, D, E and K and vitamin B12 – and some minerals, such as iron, are stored in the liver cells, until needed

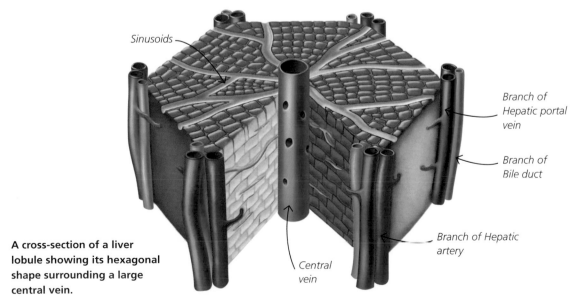

Sinusoids

Branch of Hepatic portal vein

Branch of Bile duct

Branch of Hepatic artery

Central vein

A cross-section of a liver lobule showing its hexagonal shape surrounding a large central vein.

Other Jobs for the Liver

The liver is one of the busiest organs in the body. Its other roles include:

- **Detoxification of poisons**
  Some toxic substances, such as alcohol and drugs, are removed by the hepatocytes for detoxification purposes.

- **Defence and recycling**
  Blood cells, that have come to the end of their life are destroyed by specialized phagocytic (Kupffer) white blood cells, which also destroy bacteria present in the blood sinusoids. The haemoglobin from red cells is recycled to make bile pigments (bilirubin and biliverdin).

- **Body temperature regulation**
  Because of the numerous activities that the liver performs, this organ is a source of body heat, which is important in the body temperature regulation.

The Liver and Alcohol

Alcohol undergoes detoxification in the liver. A liver enzyme (alcohol dehydrogenase) speeds up the detoxification process that converts alcohol to acetaldehyde – an intermediary, which is used to produce energy (ATP).

Individuals vary in their ability to detoxify alcohol, and alcoholics can detoxify more since continued exposure accelerates the enzymes involved in chemical breakdown. Remember though, it is the water that is doing the breakdown, and enzymes speed up hydrolysis, so now you have got no excuse not to drink water. However, continued over-consumption of alcohol damages cellular organelles; the resultant fibrosis impedes the passage of substances between blood and the liver cells. This damage is reversible at first, if alcohol is removed. But if over-drinking continues, the person may develop cirrhosis (irreversible scarring leading to portal hypertension, and loss of hepatocytes) and/or jaundice can occur, both of which may be fatal.

The liver has amazing powers to repair itself or carry on with its functions, even if it is badly damaged. In fact, it will continue to function after 70 per cent of its mass has been removed or destroyed. However, if enough cells are damaged and replaced by scar tissues, as has happened with this liver (above), then cirrhosis occurs, leading to liver failure.

Chapter 9
THE URINARY SYSTEM

Filtering the Blood

The production of urine begins with a filtration of excess fluids, salts and soluble wastes from the tiny blood vessels in the kidneys. This filtrate is then "concentrated" in the kidney to form urine.

Your urine trickles from the kidneys into the ureters, which take it to the bladder. Here, your urine is stored, before being pushed through the urethra and out into the toilet. The removal of urine from the body, termed excretion, helps to maintain the content of body fluids so that cells can remain healthy.

Urine

Urine is about 95 per cent water. The other chemicals removed found in the urine can be put into two main groups:

• **Waste chemicals** are substances that cannot be used in the body, such as water-soluble waste. Examples of these substances are urea (a by-product of protein breakdown in the liver), creatinine (a product produced during muscle contraction) and uric acid (a product produced by the recycling of old nucleic acids – DNA and RNA). Wastes also include the excess of chemicals, which cannot be stored in the body, as in the case of water, the main constituent of urine, as well as excess salts, such as sodium, potassium, chloride, phosphate and sulphate. If these chemicals were allowed to build up, they could disrupt the activity of body cells and upset the body's health and well-being. The term "waste" might also be used to describe substances that may be important in the body's activities, but only when they are needed, and therefore must be removed once their action is complete. Hormones, for example, must be removed from the body, or at least be inactivated, to stop their activities continuing longer than is required.

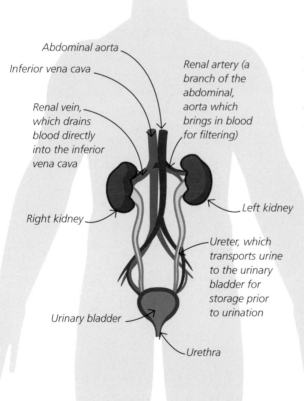

Abdominal aorta

Inferior vena cava

Renal artery (a branch of the abdominal, aorta which brings in blood for filtering)

Renal vein, which drains blood directly into the inferior vena cava

Right kidney

Left kidney

Ureter, which transports urine to the urinary bladder for storage prior to urination

Urinary bladder

Urethra

The kidneys are on either side of the back, just below the rib cage, and are linked by the ureters to the bladder.

The Urinary System

An example of this is insulin being released when blood sugar is too high, since this hormone lowers blood sugar to within its normal range. However, if this hormonal action was allowed to continue, rather than being inactivated when blood sugar levels are normal, then its activity would cause the blood sugar to get lower and lower. This could lead to symptoms such as dizziness, headaches, nervousness or shakiness, to mention just a few.

• **Any dietary substances** that cannot be stored in the body are also removed in the urine. These include water-soluble vitamins, such as vitamin C. Urine also contains many food additives that have no value to body function.

Changes in Urine

Your urine is normally sterile and varies in colour from a pale yellow, straw-like colour to a dark orange-yellow one, depending on your level of hydration. It may also be affected by pigments in food. For example, it may appear a reddish colour after you have eaten a lot of beetroot. Urine has a faint smell that becomes stronger the more concentrated it is and if it is allowed to stand. Women may notice that the odour of their urine varies with the time of the month due to the presence of female hormonal (oestrogen and progesterone) breakdown products.

In general, the kidneys are the main organs involved in removing waste substances. However the bowel, the lungs and the skin also remove unwanted and potentially dangerous chemicals from the body.

Filtration Units

The kidneys are two bean-shaped organs lying on either side of your back. The right kidney lies underneath the liver, the largest gland in the body, which displaces it and explains why it is smaller. It is also placed slightly lower in the abdomen than the left kidney. These twin organs remove waste

The odour of urine may be changed by products in food or drink, for example the distinct smell after eating asparagus or drinking alcohol.

material and material that cannot be stored in the body from the blood entering them via the renal artery, so that when the blood leaves the kidneys via the renal vein, it does not contain these chemicals. The term "renal" is from "renes", the latin for kidneys.

The Kidneys

Urine formation begins with the filtering of blood plasma. The main filtering units of the kidney are called glomeruli, and the filtered fluid (called filtrate) passes into the kidney tubules, or nephrons, that are responsible for producing your urine.

Each glomerulus is simply a tuft of blood capillaries (derived from branches of the renal artery) that provide a large surface area for filtration. The filtrate from each glomerulus moves into the individual cup-like receptacle of each nephron, called the Bowman's capsule, from which the other parts of the nephron extend. Tiny openings (pores) in the glomerulus and capsule allow the passage of smaller soluble substances, such as glucose, sodium and water-soluble vitamins, but prevent

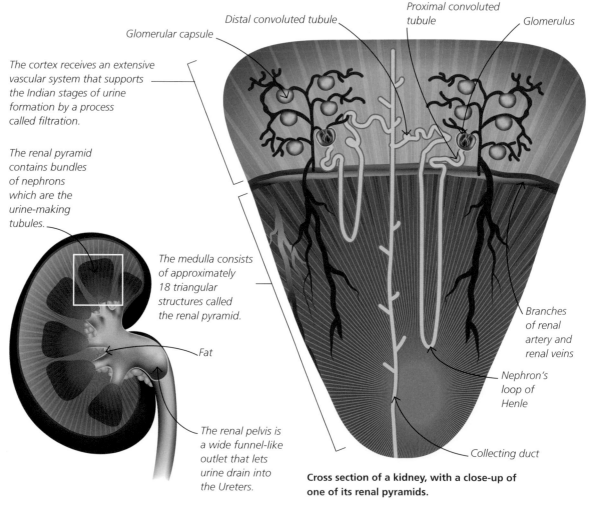

Glomerular capsule

Distal convoluted tubule

Proximal convoluted tubule

Glomerulus

The cortex receives an extensive vascular system that supports the Indian stages of urine formation by a process called filtration.

The renal pyramid contains bundles of nephrons which are the urine-making tubules.

The medulla consists of approximately 18 triangular structures called the renal pyramid.

Fat

Branches of renal artery and renal veins

Nephron's loop of Henle

The renal pelvis is a wide funnel-like outlet that lets urine drain into the Ureters.

Collecting duct

Cross section of a kidney, with a close-up of one of its renal pyramids.

The glomerulus and Bowman's capsule of the nephron, where filtrate passes from the blood into the nephron.

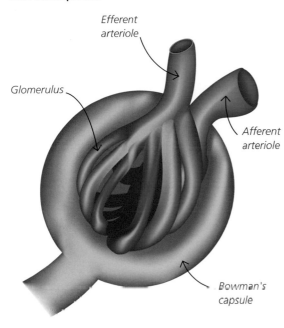

Efferent arteriole

Glomerulus

Afferent arteriole

Bowman's capsule

the passage of large materials, including plasma proteins and blood cells.

Filtering Out Waste

The chemical nature of the filtrate changes as it passes along different parts of the nephron – the proximal and distal convoluted tubules and the loop of Henle – due to a process referred to as "reabsorption". This allows the movement of dissolved substances (solutes) and water out of the nephron and back into the circulatory system. Some substances, such as sodium chloride, always undergo a degree of reabsorption, and so they are always present in the final filtrate called urine, but the rate at which they are excreted will always be less than that at which they were filtered. The kidneys play an important role in altering the rate of reabsorption of these substances. In contrast, the reabsorption of some other chemicals in the filtrate, such as amino acids and small proteins, is almost complete. They are therefore conserved in the blood and are normally absent from urine, unless there is some kidney damage.

The nephron is therefore concerned with converting the filtrate into urine, and its terminal part, the collecting ducts, finalize the concentration of urine. These ducts are influenced by antidiuretic hormone, which causes them to reabsorb more water into the blood, resulting in a small volume of concentrated urine being produced.

The final urine volume and composition vary considerably according to your state of hydration. This is dependent on many factors, including environmental temperature, activity levels, diet and water intake. However, once formed, the urine drains out of each kidney into the ureters and on to the urinary bladder.

Secrets About the Kidneys

- Kidneys are no bigger than a standard computer mouse.

- Λ third of the blood pumped out by your heart is sent straight to your kidneys, so it goes without saying that most waste is reabsorbed back into blood. This high volume is necessary because wastes, such as urea, are presented in relatively low concentrations.

- The surface area for filtration, provided by the glomeruli of both kidneys, is about the surface of a normal-sized domestic bath.

- The kidneys also activate vitamin D in your body, but only as a last resort. If your skin cells cannot produce vitamin D from sunlight, then the liver takes over. And if your liver cannot produce vitamin D, your kidneys get the job done.

- There is about a 50 per cent reduction in the volume of urine reaching your bladder at night. This is partly due to the effects of the release of antidiuretic hormone, which increases the permeability of the cells in the urinary collecting ducts since more of the filtrate is reabsorbed into the blood.

Dialysis and Kidney Transplantation

Dialysis is used in cases of kidney failure as an artificial means of replacing kidney function. However, the discontinuous nature of this treatment means that the precise regulation of blood content cannot be as accurate as that normally provided by the kidneys. Dialysis may be used in acute (sudden, short-lived and severe) renal failure temporarily until kidney function improves. In theory, if you have chronic, long-term renal failure, a renal transplantation is the ideal form of treatment, but the reality is that patients will have a prolonged spell of dialysis before a suitable donor becomes available, if at all.

The first successful kidney transplant was carried out by Joseph Smith and his team in Boston in 1954. Today, kidney transplantation is now a widespread intervention for failing kidneys. A single organ is usually transplanted into the groin of the recipient, since only one kidney is necessary for life and the single kidney will also eventually assume the role of two. In fact, 75 per cent of a properly functioning kidney is sufficient to sustain life comfortably. The transplanted kidney retains a renal artery, a renal vein and a ureter, but will have had its nerve supplies cut. So how exactly, does

a transplant work? A transplanted kidney will respond to hormones produced by the recipient and so will be able to help maintain the correct balance of body solute and fluid levels. Had renal nerves been central to regulating kidney mechanisms, then transplantation would be more problematic. Fortunately, however, hormones are the main coordinators of renal functioning.

The transplanted organ will not get back normal function immediately, and the recipient will require monitoring for signs of abnormality. Of course, there is always the possibility that the recipient will immunologically destroy and reject the kidney. For this reason, immunosuppressive drugs are administered to the recipient until the clinical team is happy that the transplant has been a success.

Ureters

The ureters are thin muscular tubes, about 30 cm (12 in) long – the length of a ruler – and one connects each kidney to the urinary bladder. The ureter is mainly made up of smooth (involuntary) muscle, and so your urine passes along them partly through peristaltic contractions and partly assisted by gravity.

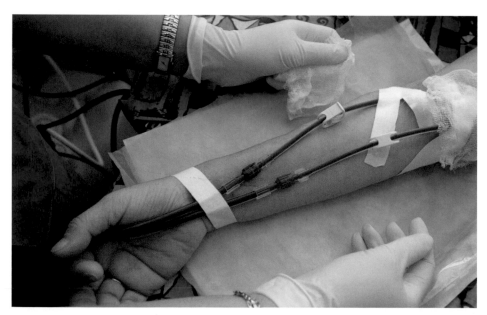

Dialysis involves taking blood from a vein in the arm for artificial filtration before returning it to the body through another arm vein.

Kidney Stones

Usually, kidney stones are mainly made up of calcium and uric acid because they have a low solubility, which means that they do not readily dissolve in water. If you live in a hard water area, you only need to look inside your kettle or around your shower head to see how calcium can build up. Deposits take place when calcium and uric acid concentrations in urine are very high.

Stones may develop anywhere in the urinary tract and result in a blockage of the renal pelvis, or nephron itself. This will lead to the holding back of urine (urine retention), the swelling of the affected kidney and substantial pain. Stones may also form in the ureters, and even within the bladder, which interferes with its emptying. The largest kidney stone ever recorded was the size of a coconut and weighed 1.1 kg (2.4 lb). Stones may be broken up using therapeutic ultrasound or removed by surgery.

Kidney stones are formed from calcium deposits and can block parts of the urinary system.

The Bladder

The bladder is a small muscular hollow organ with a "distensible" lining, which allows it to stretch when it expands as the bladder fills with urine.

The capacity of the stretch of the bladder is increased by a series of folds (called rugae and spelled the same as the folds as in the stomach). The bladder, when empty, looks like a deflated balloon, and turns pear-shaped when full of urine.

The passing of urine is referred to as micturition or urination (or simply "peeing"), and it results from an interplay of the involuntary (autonomic) nervous control of the muscle wall of the bladder and internal sphincter, as well as voluntary (somatic) nervous control of the external sphincter.

The bladder has a smooth, involuntarily controlled, muscular wall. When the bladder is full, the stretch triggers urination. The average capacity of the bladder of an adult is about 600–700 ml

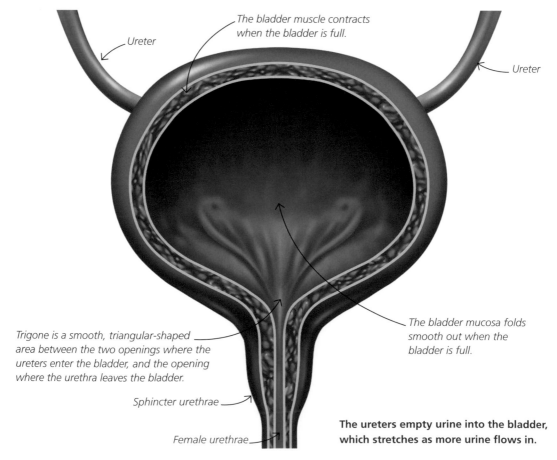

Ureter

The bladder muscle contracts when the bladder is full.

Ureter

The bladder mucosa folds smooth out when the bladder is full.

Trigone is a smooth, triangular-shaped area between the two openings where the ureters enter the bladder, and the opening where the urethra leaves the bladder.

Sphincter urethrae

Female urethrae

The ureters empty urine into the bladder, which stretches as more urine flows in.

When full, the bladder sends powerful signals, telling you to find a toilet, quickly!

controlled by the autonomic nervous system, and therefore, we generally have little voluntary control over it. The external sphincter, however, is under voluntary control – but only once it has been learnt. The development of the ability to control the external sphincter takes place in young children, and is encouraged by learning through "potty training". The need for maturation of the brain areas involved means that control is not normally established until the child is about 1½–2 years of age.

Urethra

The urethra is a tube that extends from the neck of the bladder to where it opens to the outside of the body. Urine can enter and pass through the urethra only when the internal and external sphincters of the bladder have been relaxed. In the male, the urethra is about 20 cm (8 in) long, it travels through the penis, opening at its tip. Since it is part of the urogenital system, the male urethra provides an exit from the body for urine as well as semen during ejaculation. In females, the urethra is shorter at about 4 cm (1½ in) long and emerges in front of the vaginal opening. It only provides an excretory function, since it is not part of the female genitalia.

(21–25 fl oz) of urine; this is when the very tense bladder wall activates not only stretch receptors, but also pain receptors. You may even find yourself intermittently and seemingly uncontrollably crossing your legs – this is when you need to find the toilet quickly!

Emptying the bladder requires the relaxation of two sphincter muscles. The internal sphincter is

Secrets About the Bladder

- The average bladder can comfortably hold about 0.5 litres (almost a pint) of urine, or the equivalent of two cups of urine, without you being consciously aware of its presence in the bladder.

- The scientifically estimated amount of urine that we need to pass in order to excrete all of our daily bodily toxic waste is 440 ml (15 fl oz) of urine per day. This is known as the obligatory water loss.

- It is quite normal for you to pass urine every 3–4 hours (6–8 times per day), even if you have normal hydration.

- It is also normal to sleep through the night without getting up for a pee. Elderly people may get up once a night, but do not worry as this is still defined as quite normal. It is when you get up to urinate three or more times during the night that it might be worth seeing a doctor.

- Drinking alcoholic drinks, coffee, caffeinated tea and orange or grapefruit juice can irritate the bladder and make you need to urinate more often.

Secret Roles of the Kidneys

The kidneys, through their filtering activities, have many "knock-on" effects on many important activities in the human body.

These roles include maintaining a balance in the blood of ions such as potassium, sodium and calcium. As such, the kidneys are essential for the regulation of all nervous impulses, such as the critical ones that keep us alive. This includes those that are involved in maintaining the heart rhythm (electrocardiogram). Don't forget that all muscles require nervous impulse for the contraction but these ions are also important in the contraction itself! This includes all the muscles in the body, not just the ones found attached to your skeleton, but also the heart muscle, the gut muscles, the muscles of breathing, and so on.

In their excretion of hydrogen and bicarbonate ions, the kidneys are important in regulating the acidity and alkalinity of body fluids, hence they are involved in the correct functioning of enzymes, and therefore, all the chemical reactions in the body.

Also, if needed, in their role in vitamin D activation, which is essential in calcium absorption from the gut, the kidneys also have a role in the maintenance of bones and teeth.

The kidneys are also endocrine glands. They produce a hormone called erythropoietin when your red blood cells need to be made, and another hormone called renin, which is important in regulating blood pressure.

Many of the actions this sprinter's body is carrying out involve the kidneys, including the nerve signals controlling the muscles, the contractions of the muscles themselves, as well as her breathing rates and heart rate.

Chapter 10
THE REPRODUCTIVE SYSTEMS

Reproduction

This book, so far, has shown that all the organ systems operate to maintain the well-being of the body, a process known as homeostasis. The reproductive system is also essential in maintaining this process.

At the cell level, reproduction (known as mitosis) may be regarded as a homeostatic process in which cells divide when they have reached their optimal size, when they need to be replaced or when they have come to the end of their life. This chapter of the book will focus on cellular reproduction as a necessity for growth, development and the specialization of human tissues as you go through your developmental milestones before birth. However, it will begin at the organism level of organization, where reproduction is regarded as being essential for the survival of the species.

Puberty Trigger

The function of reproduction is inactive until the onset of puberty, which seems to be a "trigger" for the production of male and female hormones to "kick-in" for the start and continuation of this developmental stage. The common goal of both sexes is to produce offspring, however, their activities are very different. The male activity is to produce a mass of spermatozoa on a daily basis, in a mixture called semen, which leaves the male by the process of ejaculation. The female produces and matures eggs (secondary oocytes) and usually releases (ovulates) one egg per month. Sometimes, though, one from each ovary is released. Following sexual intercourse, the sperm and egg may fuse in a process called fertilization to produce the offspring called the zygote. The female and male are equal partners in the fertilization process. After that, however, it is the female who takes over, providing a life-support system for the baby until birth.

Sexual reproduction requires genetic information from both male and female parents to produce offspring.

The Male Reproductive System

The male reproductive system is adapted for the production of sperm, its transport to the female reproductive tract, and producing male hormones (androgens) that regulate the changes associated with puberty.

The penis, scrotum and testicles make up the male external genitalia. The penis consists of an attached root and a free shaft that ends in an enlarged sensitive tip, called the glans penis. The shaft is covered in a loose sleeve of thin, hairless skin containing muscle fibres, which folds over to form the foreskin. The foreskin is attached to the glans penis on the underside to form a ridge of skin, the frenulum, which contains a small artery. The foreskin keeps the glans penis moist and sensitive.

Inside the Penis

The penis holds the urethra and three masses of spongy erectile tissue. During sexual arousal, the blood spaces within this spongy tissue widen and are filled with blood. The expansion squeezes the veins draining the penis, so most of the blood flowing into it stays there. This enlarges the penis causing it to become rigid. This erection allows the cylindrical shaped penis to enter the opening of the female reproductive system, so that sperm can be ejaculated inside the woman during intercourse.

The penis is also involved in transporting urine out of the body through the urethra. Ejaculation of semen is an involuntary (autonomic) nerve reflex, which causes the exit of the bladder to close. This action stops urine and semen from mixing in the urethra, which would kill sperm, and it also stops semen from entering the bladder.

The scrotum is a loose pouch of skin that is more wrinkled and darker than other body skin and often has a reddish colour. An internal membrane separates the scrotum into two compartments, each of which contains a testis (testicle).

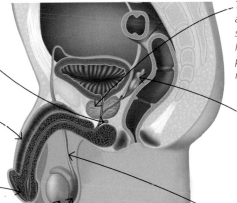

Cowper's Glands release thick, clear, alkaline mucus just before ejaculation. This neutralizes traces of acidic urine that may be in the urethra, and lubricates the end of the penis before and during sexual intercourse.

The penis is the organ of copulation and erection. During sexual arousal, the penis swells and lengthens. The opening of the male urethra is usually at the tip of the glans penis.

The glans penis is the most sensitive part of the penis.

The testis produces sperm and sex hormones.

The prostate releases a thin, milky, alkaline secretion, which liquefies semen and promotes maximum motility of the sperm.

Seminal vesicles contain a sugar that provides the essential fuel necessary to maintain the beating of the sperm tail.

The vas deferens transports sperm to the urethra.

The epididymis stores the immature sperm produced by the testis, until it is released into the vas deferens or recycled.

The male reproductive system is made up of the two testes and the tubes that carry sperm into the penis.

The Testes

The testes produce the sperm component of semen. They hang in the scrotal sac outside of the body where they are slightly cooler than the rest of the body. This is vital for normal production of sperm.

The testes store the sperm until they are either released from the body or broken down and then recycled. The testicles also have endocrine tissue, which produces and releases the male hormones. Other reproductive organs mature and protect the sperm and help them on their way out of the body. These organs include transport tubes – the vas deferens and the urethra – and various secretory glands, including the prostate gland, seminal vesicle glands and Cowper's glands. These glands produce the chemical secretions of semen.

Producing Sperm

The paired, oval-shaped testicles are divided inside into lobules. Each lobule contains the convoluted seminiferous tubules, which produce sperm. Embedded between the developing spermatozoa cells are specialized cells, and these support, nourish and protect developing sperm cells.

Inside the testis, the seminiferous tubules carry the sperm to the epididymis where they are stored prior to ejaculation.

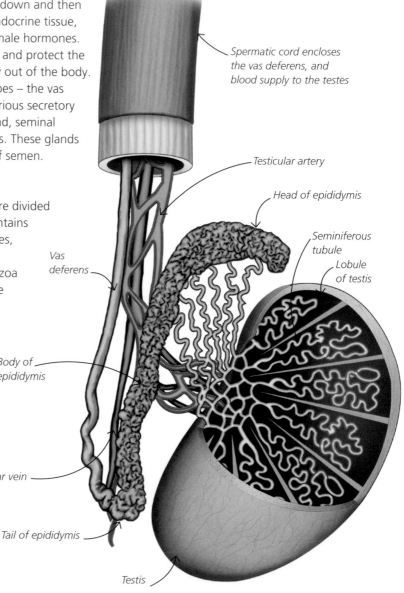

Spermatic cord encloses the vas deferens, and blood supply to the testes

Testicular artery

Head of epididymis

Seminiferous tubule

Lobule of testis

Vas deferens

Body of epididymis

Testicular vein

Tail of epididymis

Testis

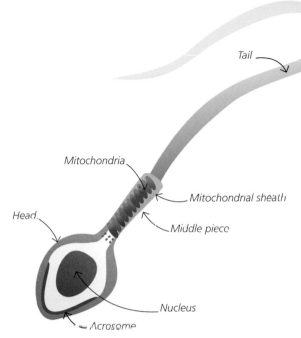

Each sperm has a tail-like projection that propels it. The sperm's nucleus contains the father's genetic material and its middle-piece has many mitochondria, to provide the energy for the movement of the tail.

The process of sperm production by the tubules is regulated by the release of the follicle-stimulating hormone (FSH) from the brain's anterior pituitary gland. The testes' release of the male hormones (androgens) is controlled by luteinising hormone (LH) secreted from the anterior pituitary. Testosterone is the main androgen produced by the testes, and this is important in the development of the secondary sexual characteristics at puberty

After production, the spermatozoa are transported through a duct system that is 8 metres long! On their way out of the penis, sperm pass through the epididymis, the vas deferens (sperm tube) and the urethra. The epididymis is a coiled tube attached to the back of the testes. Here, the sperm's tail develops its swimming capability and the sperm becomes fertile. The epididymis is also a temporary storage site for sperm until they are either broken down chemically, with their constituents being

recycled, or released into the vas deferens when the male is sexually aroused and then ejaculates.

The sperm tube, or vas deferens, starts from the epididymis and terminates just behind the bladder. Each sperm tube empties into the ejaculatory ducts, the next part of the male reproductive system. The muscular walls of these ducts contract and empty their contents into the urethra.

Ejaculation occurs when peristaltic contractions from the testes spread to the vas deferens and accessory glands – the prostate, seminal vesicles and Cowper's glands. At the same time, the exit to the bladder closes. Muscles in the penis contract, and the semen is pushed out through the urethra and expelled from the body. The secretions from the accessory glands activate spermatozoa in the semen by providing nutrients for their motility. They also counteract the acidic environment of the female reproductive tract.

Gametogenesis

Gametogenesis is the name given to the production of the male and female gametes (sperm and ova, also called oocytes). The gametes are produced from a specialized type of cell division called meiosis, covered in chapter 1. Basically, this type of cell division ensures that the sperm and ova contain half the chromosome number of the parent cell from which they originate. That is, gametes have 23 chromosomes, while parent cells have 23 pairs. This type of cell ensures that when the sperm and egg come together at fertilization, the normal number of chromosomes is restored to 23 pairs in the fertilized egg, called the zygote. Furthermore, since chromosomes are made of genes, each gamete when formed will contain a random half-selection of genes present in the parent cell. Put simply, this is why we resemble our parents and our brothers and sisters because this selection of mum and dad genes are mixed together to produce you at fertilization.

Secrets About the Male Reproductive System

- In circumcision, the foreskin is surgically removed, causing the skin of the glans penis to lose its soft, moist texture. As fibrous protein is laid down on the glans, it becomes more like normal skin and some sexual sensitivity may be lost.

- Follicle-stimulating hormone and luteinizing hormone are named after their role in stimulating ovarian follicles in women and to produce the corpus luteum, respectively, but they have retained their names in men.

- Production of sperm starts at puberty and continues throughout a man's life, whereas the production of ova starts at puberty but stops at menopause.

- A vasectomy involves cutting or tying the vas deferens. The testes are still able to produce sperm, but it is now unable to be discharged outside of the body. It eventually dies and its components are recycled by nearby cells. However, although the man becomes infertile by this procedure, he is still able to ejaculate the secretions made by the accessory glands during sexual intercourse.

Divided Vas deferens

Epididymis

Testis

A vasectomy involves the cutting and/or tying of the vas deferens leading from both of the testes, preventing passage of the sperm to the penis.

Sperm are produced by the testes well into old age.

Testicular and Prostatic Cancers

Although rare, testicular cancer is 20 times greater in people with a history of late descended or undescended testes. Most testicular cancers arise from the sperm-producing cells.

Testicular tumours represent about 2 per cent of male cancerous growth (malignancies), and it is the most common form of cancer affecting men in their early twenties.

Because of this risk, health care practitioners, such as nurses and doctors, should encourage frequent self-examination of the testicles, since an alteration of or an enlarged testicle should be reported to a doctor as soon as possible, even though it may not necessarily be a malignancy.

Prostate Cancer

The prostate is the only organ or gland in the body that grows with age, and this growth may impede the flow of urine and produce pain during urination. The prostate lies just below the bladder and just in front of the rectum, and a doctor can feel its size and texture by a digital (finger) examination through the wall of the rectum.

Prostatic cancer is a frequent and leading cause of cancerous death in men. However, it rarely occurs before the age of 50 years. Genetics is a known risk factor and, therefore, the risk is greater if a close relative has or had prostate cancer. There also seems to be a link if several female family members have developed breast cancer, especially if it is present under 40 years of age, indicating there may be a common genetic link in these cancers.

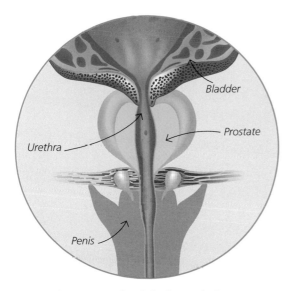

A "normal" prostate gland sits beneath the bladder and around the urethra, without reducing its diameter.

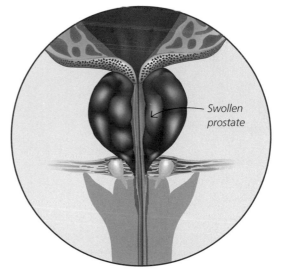

During prostate cancer, the swollen prostate gland squeezes on the urethra, reducing its diameter and slowing the flow of urine out of the body.

The Female Reproductive System

The female reproductive system consists of the clitoris, vagina, cervix, uterus, Fallopian tubes and ovaries. It not only produces ova, but also provides a "home" for the developing offspring.

The female reproductive system has internal and external sex organs. The external genitalia (vulva) consists of the clitoris, labia majora and labia minora ("labia" meaning "lips"). In adult females, the vulva is covered by pubic hair, which helps to regulate airflow and warmth around the genitals, and traps pheromones – chemicals involved in sexual attraction. The clitoris is the female equivalent of the glans penis and it has a similar structure. The labia minora are a pair of thin, red skin folds that surround the entrance to the vagina.

They vary in size and shape and one is often longer than the other. The labia majora surround the labia minora on each side.

The Hymen

The hymen is a thin membrane that partly closes the entrance to the vagina in young girls and women who have not had sex. It allows discharge and menstrual flow to pass through. The hymen often breaks down normally after puberty, when playing sports, having sex or inserting a tampon.

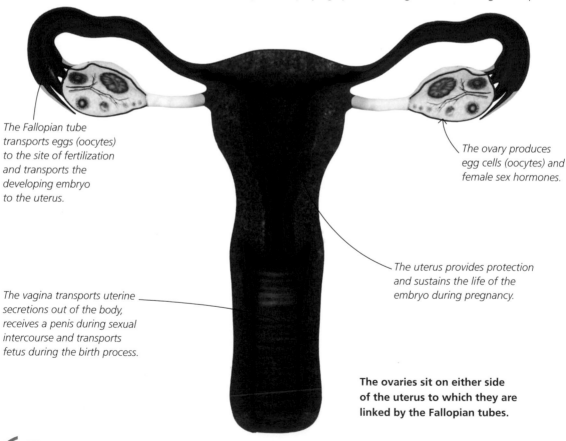

The Fallopian tube transports eggs (oocytes) to the site of fertilization and transports the developing embryo to the uterus.

The ovary produces egg cells (oocytes) and female sex hormones.

The vagina transports uterine secretions out of the body, receives a penis during sexual intercourse and transports fetus during the birth process.

The uterus provides protection and sustains the life of the embryo during pregnancy.

The ovaries sit on either side of the uterus to which they are linked by the Fallopian tubes.

Internal organs

The internal organs are the vagina, uterus, Fallopian tubes and ovaries.

The vagina connects the vulva to the upper reproductive tract. It is a distensible organ that acts as a passageway for blood loss during menstruation and for childbirth, which is why it is often referred to as the birth canal. The vagina is also the receptacle for the penis and semen during intercourse. Because of this, its wall consists mainly of involuntary muscle and its folded lining is lubricated with mucus to facilitate intercourse.

The vaginal secretions are acidic and, as such, provide a hostile environment to protect against any potential microbes that may enter the vagina. The acidic region is also hostile to spermatozoa. However, the alkaline nature of semen neutralizes this acidity, thus ensuring sperm's survival. But this is only partly successful, since most sperm die before the neutralizing process becomes effective.

The Uterus

The uterus is a hollow, muscular, upside-down pear-shaped organ, which consists of the dome-shaped fundus region, the main body region and the neck of the uterus called the cervix, which continues as the vagina below the uterus. Its wall has an outer layer (perimetrium), a thick muscle layer (myometrium), and the inner endometrium. Functionally, the endometrium has two layers:

- **The inner functional layer**, which becomes thickened and rich in its blood supply in the first half of the menstrual cycle. If fertilization does not take place, part of this layer is removed and shed through the opening of the vagina during menstruation.
- **The outer basal layer**, which is not shed during menstruation, but is concerned with the replacement of the functional endometrium in the second half of the menstrual cycle.

The uterus is the site of implantation, embryo and foetal development and is where labour occurs. During this time, the uterus changes in size dramatically, from its "normal" non-pregnant size,

gradually increasing in size during pregnancy to house the growing foetus to its maximum size just before birth. The neck of the uterus, the cervix, opens into the vagina. The cervix closes to accommodate the embryo/foetus in the uterus until birth. During labour, it widens, to allow the passage of the foetus out to the vagina.

The Ovaries

The paired ovaries are about double the size of an almond nut and, like the testes, are found close to the posterior abdominal wall near the developing kidneys. During embryological development, they descend to just below the pelvic brim, where they remain.

The ovaries are held in position by ligaments, which attach them to either side of the uterus and to the pelvic wall. The ovaries produce the sperm's "sister", the oocytes, (cells that develop into a mature egg or ova). They also produce female hormones (progesterone, oestrogens), which regulate the growth of tissues within the endometrium and the breast to prepare for implantation and milk production (lactation) following birth.

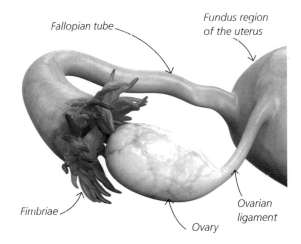

Fallopian tube

Fundus region of the uterus

Fimbriae

Ovarian ligament

Ovary

The fimbriae are finger-like projections sticking out from the entrance to each Fallopian tube. At ovulation, they act as guides to make sure the released egg enters the Fallopian tube.

The Menstrual Cycle

The menstrual cycle involves cyclical changes in the lining of the uterus (endometrium) and breast in a non-pregnant woman in response to changing levels of hormones released by the ovaries.

Each month, the endometrium is prepared so that it's ready to receive a fertilized egg. This preparation is essential for embryo and foetal development during pregnancy. If fertilization does not occur, however, part of the endometrium is shed as the menstrual flow. The ovarian cycle involves changes, such as the maturation and ovulation of the female egg (technically called the secondary oocyte) that occur in the ovaries during the menstrual cycle. The control of both cycles is regulated by the gonadotropin hormones (FSH and LH) from the brain's anterior pituitary gland, and the female sex hormones, oestrogen and progesterone from the ovary.

The menstrual cycle usually has a duration of 28 days, however, it can be shorter or longer depending on the individual's response to environmental factors, such as stress and infection, and will cease completely at menopause. The cycle begins with the first signs of menstrual bleeding (called menses). This occurs when the endometrium, along with the enhanced blood supply that developed in the preceding cycle, begins to be shed. This usually lasts for around four to five days. Prior to and during menses, the release of oestrogen and progesterone is decreased, and it is the lack of these that stimulates the shedding of the endometrium. As these ovarian hormone secretions are reduced, they stimulate an increased secretion of the gonadotropins (FSH, LH) from the anterior pituitary, and this, in turn, stimulates development of the next cycle. The release of the mature female egg (secondary oocyte) is called ovulation, and this usually occurs midway through the menstrual cycle.

Ovarian Problems

Ovarian cysts are fluid-filled sacs, which are normally benign (non-cancerous) and rarely

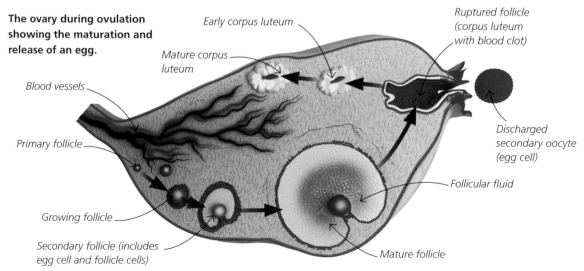

The ovary during ovulation showing the maturation and release of an egg.

Early corpus luteum

Mature corpus luteum

Ruptured follicle (corpus luteum with blood clot)

Blood vessels

Primary follicle

Discharged secondary oocyte (egg cell)

Follicular fluid

Growing follicle

Secondary follicle (includes egg cell and follicle cells)

Mature follicle

The Reproductive System

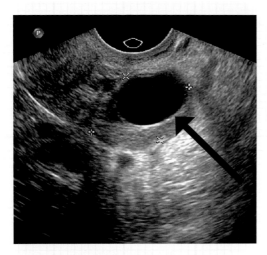

This ultrasound scan shows an ovarian cyst (the black space, arrowed). Since the majority of ovarian cysts cause no symptoms, many are diagnosed by chance, for example, by a routine examination.

- Pap smears detect early changes to the cervical cells which may lead to cervical cancer. However, the vast majority of abnormal Pap smears are due to inflammation or infection.

- Every month, between 100–150 eggs begin to mature (under the influence of FSH) inside the ovaries, although only one reaches full maturity.

- If the egg is fertilized, pregnancy occurs and the developing placenta releases a hormone called human chorionic gonadotropin (HCG). The presence of this hormone gives a positive result in the pregnancy test.

- Most women are completely unaware of ovulation. However, some may experience low abdominal pain, usually on the side near the ovary that is ovulating. This pain is due to the pressure within the swollen ovarian follicle.

- Usually, only one egg is released each month. However, in the case of twins, two eggs are ovulated at the same time, one from each ovary.

- Eggs are not released alternatively from each ovary, but they are released from either ovary in an irregular and unpredictable pattern.

become dangerous. They frequently disappear within a few months of their emergence. However, some are cancerous, or may become cancerous over time. Two of the more common ovarian cysts are follicular and corpus luteum cysts. The cyst develops when there is a "fault" during ovulation. Follicular cysts develop from a follicle that fails to rupture completely at ovulation, while corpus luteum cysts develop from the corpus luteum if it fails to degenerate in the second half of the cycle. The cysts vary in size, from less than the size of a baked bean to the size of a tennis ball. Follicular cysts are the smaller of the two types of cyst. Only a minority of cysts cause problems, such as constant or intermittent pain in the lower abdomen and irregular, heavy or lighter periods.

Polycystic ovarian syndrome translates as "many cysts" in the ovary. These cysts develop due to a problem with ovulation caused by a hormone imbalance. However, the exact cause is not clear, and glandular cells within the cysts themselves cause further hormone imbalance. This syndrome is associated with period problems, such as reduced fertility, hair growth, obesity and acne.

About one birth in every 250 is a pair of identical twins.

Conception

Following ovulation, the secondary oocyte is collected by the funnel-shaped ends of the Fallopian tubes. The oocyte is then transported to the uterus (womb), during which time it may be fertilized.

The Fallopian tubes have a ciliated lining, and these cilia pulse to help the movement of the oocyte along the tube. There is also a muscular layer in each Fallopian tube which contracts in waves to create peristalsis that also helps to move the oocyte.

This is where the terminology becomes tricky! The secondary oocyte becomes an ovum if it is fertilized by a sperm. So an ovum is when the sperm and oocyte nuclei (containing the father's and mother's chromosomes) are still separate. When the nuclei join and the parental

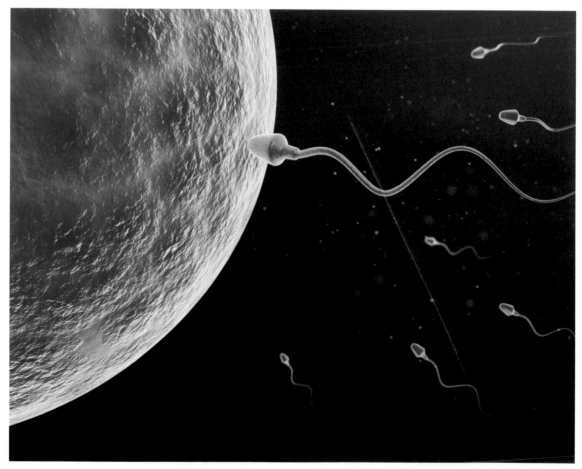

Sperm are hundreds of time smaller than an oocyte and only one will complete its journey to fertilize the oocyte (egg).

The Reproductive System

chromosomes come together, this one-cell product of fertilization is now called the zygote. This cells develops into the embryo, which must implant in the endometrium for a pregnancy to occur.

The hormonal control of the menstrual and ovarian cycles must promote endometrium growth, spur on ovulation at a time when the endometrium is ready for implantation to occur, promote further nutrient support of the endometrium and prevent its shedding. These will help to sustain the embryo should implantation occur, and cause shedding of the endometrium if conception (fertilization) does not take place.

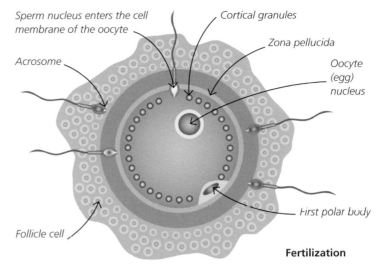

Fertilization

In order to fertilize an oocyte and complete fertilization, a single sperm has to burrow its way through protective follicle cells, before working its way through the oocyte's own cell membrane and releasing its genetic information into the egg.

Secrets About Sperm, Ova and Fertilization

- **Did you know that sperm production starts at puberty and continues throughout life, whereas in ova production, all the immature eggs, about 750,000, are present before birth?**

- **Maturation of eggs is stimulated by hormonal changes following puberty.**

- **Once the menstrual cycle is established, only about 500 immature eggs, called primary oocytes, begin their developmental cycles each month. The remainder are broken down and their component chemicals recycled for other ovarian cells to use.**

- **Usually, once a sperm nucleus has penetrated the membrane of the mature egg, a chemically induced fertilization membrane is set up, removing the "attracting powers" of the secondary oocyte, preventing the entrance of other sperm. Sperm are not interested in two-timing!**

- **Non-identical twins result from two eggs being fertilized, and simultaneous ovulation from both ovaries must occur for this to happen.**

- **On the other hand, identical twins result from the splitting of cells from one fertilized egg.**

- **Fertilization by more than one sperm is called polyspermy and this inevitably leads to early embryonic death.**

- **Conception technically starts when the sperm fuses with a secondary oocyte and it is completed when implantation occurs and the placenta starts to develop.**

Development in the Womb

The zygote divides repeatedly every 11 hours, eventually producing a protective encapsulated ball of 64 cells (the morula), which is moved along in fluid within the Fallopian tube to the uterus over a few days.

The cells arising from the zygote have an identical number of chromosomes (23 pairs). However, by the time the morula reaches the uterus, the cells begin to differentiate. Some become embryonic cells, while others surround the embryo and are destined to develop into the placenta and umbilical cord. This fluid-filled ball of cells, known as the blastocyst, becomes implanted into the uterine wall (endometrium) about three days after its arrival.

The embryo cells of the implanted blastocyst develop into three layers – the ectoderm,

mesoderm and endoderm. This developmental stage sees the start of the basic body plan. The ectodermal cells become the outer layer (or epidermis) of the skin and the nervous system. The mesodermal cells start to turn themselves into the skeleton, its muscles, blood cells and most of the body's internal organs. The endodermal cells specialize into the cells of the digestive system and its associated organs. The embryo stage is concerned with this organ system development and the embryo becomes increasingly humanoid in shape. This stage is mostly complete by about two

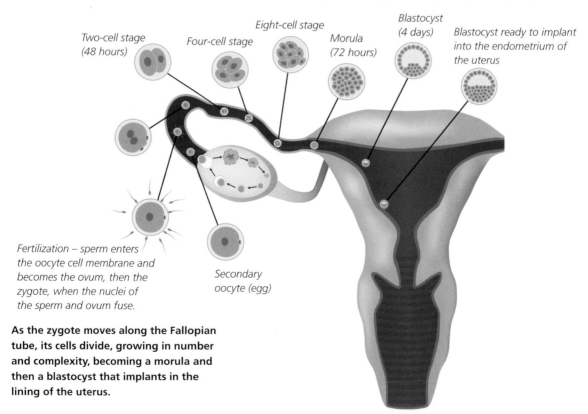

Two-cell stage
(48 hours)

Four-cell stage

Eight-cell stage

Morula
(72 hours)

Blastocyst
(4 days)

Blastocyst ready to implant
into the endometrium of
the uterus

Fertilization – sperm enters
the oocyte cell membrane and
becomes the ovum, then the
zygote, when the nuclei of
the sperm and ovum fuse.

Secondary
oocyte (egg)

As the zygote moves along the Fallopian tube, its cells divide, growing in number and complexity, becoming a morula and then a blastocyst that implants in the lining of the uterus.

A whitish coat of a slick, fatty substance called vernix caseoso begins to cover the foetus, protecting the skin during its long immersion in amniotic fluid.

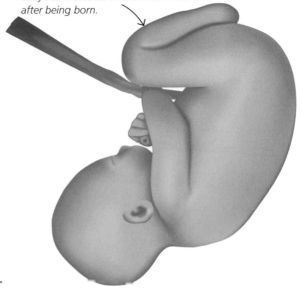

The average baby is about 51 cm (20 in) long from head to toe and weighs about 3.4 kg (7½ lb) at birth. By this time, the baby has mastered all the skills it will need after being born.

During the last stages of foetal development, the unborn child prepares itself for life outside the womb.

months after fertilization, at which time, the foetal stage begins.

During the early stages of embryonic development, the embryo is particularly susceptible to harmful environmental agents, such as drugs, certain microbes and alcohol. These are small enough to cross the selectively permeable placenta, and can cause serious birth defects. A point to note is that, unfortunately, this is also a time when the mother may not even know she is pregnant.

Further development of the foetus principally involves growth and the functional maturation of the organs that formed during the embryonic period. The foetus also increases in size by up to 1,000 times. Unfortunately, this does not mean that the foetus is now no longer at risk of defects of development, but the risks do decline.

Pregnancy

The length of pregnancy is calculated from the first day of the last blood loss (menses), however, fertilization occurs about two weeks later, following ovulation. A baby's gestational or developmental age is, therefore, two weeks less

than the calculated length of pregnancy. Because of this, when a woman is seven weeks pregnant, the gestation age of the baby is five weeks. The average gestation is 38 weeks from fertilization to childbirth, this is the same as 40 weeks of a usual pregnancy.

The Placenta

The placenta is a pancake-shaped organ, which is responsible for the exchange of materials between the foetal and maternal circulations. It is quite unique, since it links two individuals – the mother and the foetus. The placenta forms inside the mother's uterus and attaches to the foetus's umbilicus (navel) region by the umbilical cord. At the placenta, the foetal blood picks up oxygen and nutrients, and drops off carbon dioxide and other wastes into the mother's circulation. Following birth, the placenta detaches from the uterus and becomes the "afterbirth".

Trimesters

Pregnancy is divided into three periods called trimesters. The duration of the first trimester is 1 to 12 weeks, the second is 13 to 27 weeks, and the final trimester is between weeks 28 to 40.

The previous pages deal with the main changes associated with the first trimester and the transition of embryo to foetus. The placenta develops and becomes fully functional at approximately 12 weeks of pregnancy, which marks the end of the first trimester.

The second trimester marks the time of greatest activity for the foetus, because it has a lot of space in the amniotic sac to bend, stretch, make complex movements with its hands and even do somersaults. For the mother (and her partner), it is generally the most enjoyable period of pregnancy, even though these foetal movements may be quite energetic.

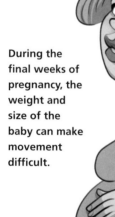

During the final weeks of pregnancy, the weight and size of the baby can make movement difficult.

Secrets About the Trimesters

- During the first trimester, the baby hiccups regularly to exercise its diaphragm muscle and glottis.

- The average maternal weight gain by the 20th week of pregnancy is 4–6 kg (9–13 lb) but this will vary between women.

- If born early, a 24-week-old foetus has a small chance of survival in an intensive care baby unit.

- The length of foetal gestation is about equal to the height of the uterus above the pubic bone measured in centimetres. So a fundal height of 26 cm (10 in) approximates to a gestational age of 26 weeks.

- The full-term for a single baby pregnancy is 40 weeks, however, a birth time between 38 to 42 weeks is accepted as "normal".

Most of the early symptoms of pregnancy have disappeared and the baby has not grown so large that he/she is uncomfortable to carry about.

During the final trimester of pregnancy, the foetus grows significantly, and more than triples in weight from about 910 g to 3.4 kg (2 to 7½ lb). The added weight of the baby and pressure on the mother's internal organs and skin can lead to several uncomfortable symptoms; frequent urination, indigestion, heartburn and stretch marks.

Childbirth

When the mother is about to give birth, she is said to "go into labour". Labour is often divided into three phases. These are known as early (dilation) labour, active (expulsion) labour and placental labour.

During labour, the hormones oxytocin and prostaglandins regulate the contractions of the uterus. When the mother is in early labour, contractions are short (lasting between 9 and 14 seconds each) and far apart (normally every 30 to 40 minutes) and irregular. As the labour advances, these contractions increase their frequency, with one every 1 to 3 minutes, and lasting about one minute as birth is about to happen.

During the early (dilation) stage, the cervix of the uterus is fully dilated to 10 cm (4 in) by the head of the foetus. The amniotic membranes rupture releasing the amniotic fluid. This is commonly referred to as "breaking the waters". Dilation is the longest stage of labour, lasting up to 12 hours.

In the active (expulsion) stage, the child moves through the cervix and the baby's head will appear at the vaginal opening, known as "crowning", and the baby is then delivered to the outside world. This stage normally lasts about 15 minutes in the first birth to about 20 minutes in future births. Normally, the head of the child moves out first, and the nose and mouth are cleared of mucus so the child can breathe. The umbilical cord is cut and clamped after the rest of the child's body has emerged. A breach birth is one in which the buttocks emerge first and delivery is more difficult, sometimes resulting in a caesarean section, where the baby is removed by cutting open the abdomen.

During the placental stage, the placenta detaches from the uterus about 12 to 15 minutes after birth. The removal of all the placental material will prevent prolonged bleeding after delivery.

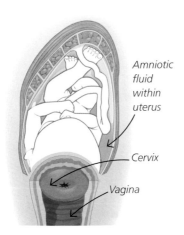

Amniotic fluid within uterus

Cervix

Vagina

Dilation
The cervix opens, or dilates to allow the baby's head to pass out of the uterus.

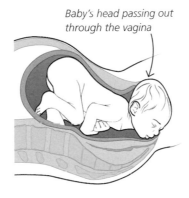

Baby's head passing out through the vagina

Presentation of head
The head passes out through the vagina, the body twists to ease the baby's passage.

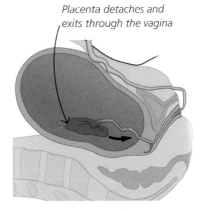

Placenta detaches and exits through the vagina

Afterbirth delivery of Placenta
About 15 minutes after the baby's birth, the placenta and its membranes pass out of the vagina.

Where Does Your Belly Button Come From?

In the absence of external interventions, the cord closes shortly following birth, in a response to a sudden reduction in temperature; this is accompanied by a constriction of the blood vessels, which in effect, acts as a natural clamp, halting the flow of blood.

Usually within two weeks of birth, the remnant of the umbilical cord that is attached to the baby starts to shrivel and falls off. The area where the cord was located becomes covered by a thin layer of skin, and scar tissue forms. The scar is called the navel or umbilicus.

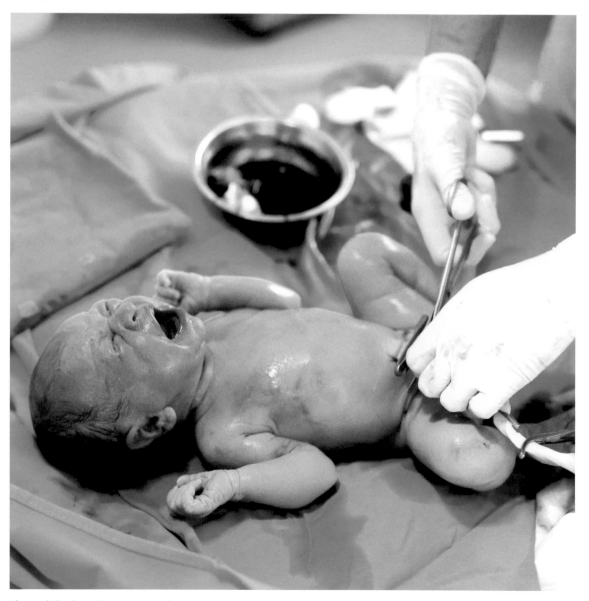

The umbilical cord is cut soon after birth, leaving a small fragment, which drops off after about two weeks.

The Reproductive System

Secrets About Babies

- Only 5 per cent of babies are born on their "due date", which is calculated at the beginning of pregnancy.

- Newborn babies need a lot of sleep. For the first few weeks, a baby sleeps for about 18 hours over the course of 24 hours.

- Did you know that the average childbirth lasts for 7 to 9 hours for the first baby, and around 4 hours for deliveries after the first-born?

- Most full-term babies weigh around 2.7 to 4.3 kg (6 to 9½ lb).

- Most babies lose between 5 to 10 per cent of their body weight in the first few days after birth. The majority of these babies get back to their birth weight by the time they are two weeks old.

- About 30 per cent of babies are born with a birthmark (often called angel kisses) and most birthmarks disappear on their own, but they're not dangerous so don't worry if they don't.

- A newborn baby grows around 2 cm (¾ in) during each of the first few months.

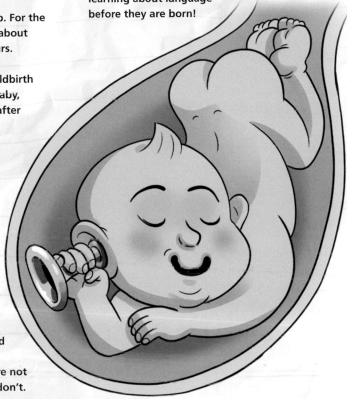

Your baby could hear your voice from about 24 weeks of pregnancy. So it could be argued that she or he is learning about language before they are born!

- A baby is born with the ability to swim as they naturally hold their breath when underwater!

The Mammary Glands

The mammary glands (breasts) are accessory organs of the reproductive system that are present in both sexes. The breasts are actually modified sweat glands and, as such, they are a part of the skin and its fat layer. The size of the breast depends on the amount of fat surrounding this glandular tissue. Each breast contains a pigmented circular region of skin, known as the areola, which surrounds a central protruding nipple. The areolar

sebaceous glands give this region a slightly bumpy appearance. These glands secrete sebum to keep the areola and nipple lubricated while breast-feeding a baby. Exposure to cold temperatures and tactile (touch) or sexual stimuli, stimulates the muscle fibres in the areola, causing the nipple to become erect.

Inside, each breast is made up of 20 to 25 irregularly shaped lobes radiating around the nipple. Each of these is separated from the others by the ligaments of the breast. In each lobe, smaller lobules called alveoli are present and the milk-secreting cells are found in these. During lactation, milk passes from the alveoli to the lactiferous ducts. The ducts enlarge to form the lactiferous sinuses just before their openings on the nipple. The milk collects in the sinuses during "nursing". In non-pregnant, non-nursing women, the breasts and the duct system are underdeveloped.

The Breasts at Puberty

When a girl approaches puberty, the breast remains underdeveloped, but a surge of ovarian hormones (oestrogen and progesterone) stimulates further development of the mammary glands. Oestrogen stimulates the development of the duct system and progesterone activates the development of the alveolar regions. In boys, secretion of the same female sex hormones from their adrenal glands may also cause the breast to appear enlarged, because of the accumulation of fatty tissue. During puberty, some males experience gynaecomastia, in which the breasts enlarge temporarily as a result of this hormonal imbalance.

Since their role is to produce and secrete milk in order to provide a source of nourishment to a newborn baby, a woman's breasts only become biologically functional following pregnancy. The regulation of breast development from puberty onwards, during pregnancy and lactation, involves the female sex hormones. However, the initiation and maintenance of lactation involves two further hormones – prolactin and oxytocin. These are released from the brain's pituitary gland, and they stimulate milk production and control milk let-down, respectively.

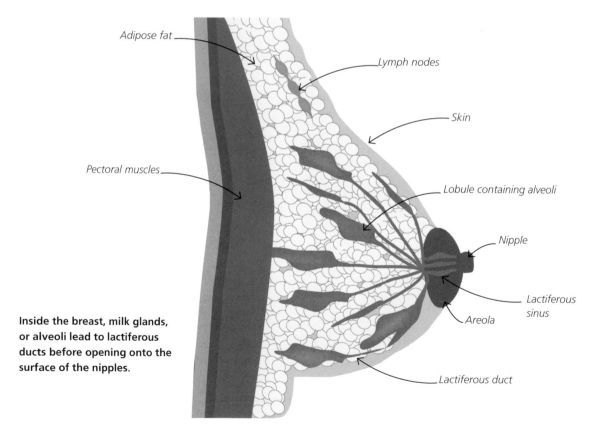

Adipose fat

Lymph nodes

Skin

Pectoral muscles

Lobule containing alveoli

Nipple

Lactiferous sinus

Areola

Lactiferous duct

Inside the breast, milk glands, or alveoli lead to lactiferous ducts before opening onto the surface of the nipples.

The Reproductive System

Chapter 11
GENETICS AND INHERITANCE

The Secret Code of Life

The human body is made up of billions of cells, each specialized for their specific activities within the body. This specialization is primarily controlled by the secret unique genetic information passed on to you by both your parents at fertilization.

The activities of genes are clearly seen throughout life in their role, for example, in reaching our developmental milestones and in identifying health parameters and not just disease. Although many of them are very rare, there are more than 3,000 genetic disorders that are passed on (or inherited) from the parents to their offspring and you will have heard of the commoner ones, such as cystic fibrosis and haemophilia.

Inherited disorders are usually present at birth, although the effects of some inherited genes may only present themselves later in childhood, or even adulthood. This is because gene expression may be "masked" for a while. For example, disorders such as cancer and Alzheimer's disease are now known to have genetic components that only become problematic later on, because the genetic code is altered during life as the individual increases their exposure to the environmental risk factors associated with these diseases. In relation to

ill health, therefore, two important features are firstly that the genes that you inherit either influence you in your embryological and foetal development or give you a susceptibility to later acquire these conditions, and secondly, the genetic changes that occur during your lifespan.

The chance of passing on genetic disorders to any offspring can be worked out using a pedigree analysis of your family history, which identifies the presence of a "diseased" gene within the family line by going back through several generations. This analysis, however, can only highlight the mathematical chances of an offspring inheriting a diseased gene, as parents may be healthy carriers of the diseased gene. Consequently, a couple often only become aware that they are healthy carriers at the birth of a child who has the genetic disease. It is only then that genetic counsellors can inform the parents regarding the chances of passing the disease gene on again in further pregnancies.

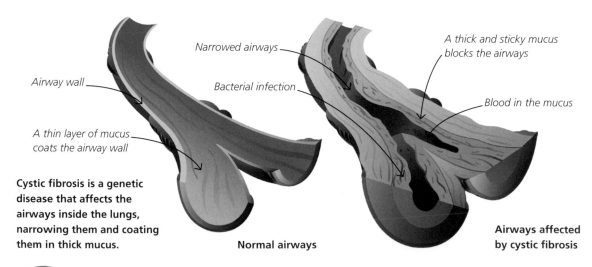

Narrowed airways

A thick and sticky mucus blocks the airways

Airway wall

Bacterial infection

Blood in the mucus

A thin layer of mucus coats the airway wall

Cystic fibrosis is a genetic disease that affects the airways inside the lungs, narrowing them and coating them in thick mucus.

Normal airways

Airways affected by cystic fibrosis

Genetics and Inheritance

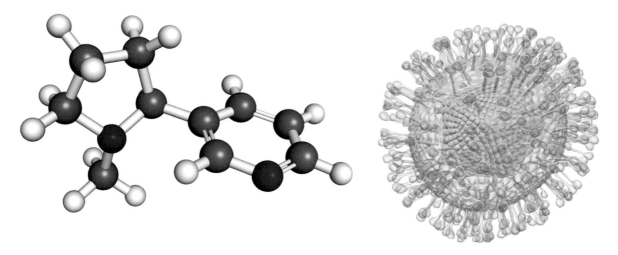

Small molecules, such as nicotine (left), and even viruses, such as chickenpox (right), can move through the placenta, passing from the mother to the unborn infant.

Nature–Nurture Interactions

Congenital disorders are present at birth These disorders affect development and biochemical activities of body parts, and it is normally the case that they have an impact on health. They arise from a failure of differentiation of tissue in the early embryo, and this is usually of a genetic origin. Alternatively, it may be a defect or failure of further development of a tissue, and this can also have a genetic cause or may be due to an environmental factor. For example, alcohol and drug abuse by the mother, or a virus, such as HIV, passing over the placenta, or a deformation occurring, such as an altered tissue size or shape in an organ arising from the effects of physical forces on tissue growth. An example of this would be the limited brain growth in a foetus with hydrocephalus, a condition where excessive cerebrospinal fluid compresses the brain tissue against the bones of the skull.

Tissue differentiation and development, therefore, largely depends upon the expression of genes and the actual genetics inherited by the embryo.

Gene Expression

Environmental agents that prevent cell specialization and differentiation in the embryo stages of development that cause physical problems in the newborn are collectively called teratogens. These can include "mutagens" that alter genetic structure, but also include those agents that alter the expression of genes, that is, switching genes "on" when they should be switched "off" and vice versa.

Teratogens, being environmental agents, have the potential to be introduced in some form into the embryo/foetus by passing through the placenta. Teratogens include maternal infective microbes such as those which cause chickenpox, hepatitis, herpes, mumps, pneumonia, polio, rubella and tuberculosis. All of these increase the risk of congenital problems arising, and some can cause foetal death. Teratogens also include numerous chemicals found in drugs such as aspirin, antihistamines, morphine, heroin, LSD, nicotine, thalidomide and alcohol. Some of these drugs may be relatively harmless to the mother, but can have devastating effects on embryonic/foetal cells. A lack of some chemicals may also be teratogenic. For example, the incidence of spina bifida can be reduced by ensuring that the mother's diet contains adequate folic acid, since this vitamin plays an important role in neural development.

Inherited Genetics

Genetics is the study of heredity, the passing of characteristics from parents to children. Physical characteristics, called phenotypes, such as hair and eye colour, are inherited, as well as biochemical and physiological traits, including the likelihood to develop genetic disorders.

Inherited characteristics are transferred from parents to offspring via the gametes, so your inheritance is set at fertilization, when an ovum and sperm are united.

In the nucleus of each gamete are the chromosomes. Each chromosome is composed of the genetic material called deoxyribonucleic acid (DNA). DNA is a long molecule made up of thousands of segments called genes. Each of the characteristics that you inherit, from the texture of your hair to a disease you were born with, are carried by your unique secret code of genes. The location of a gene on a chromosome is often likened to beads (genes) on a necklace (chromosome). The position (or locus) of each gene is specific and does not vary between people.

The genotype for red hair is recessive, while the gene for brown hair is dominant. So a recessive gene can only be passed from the mother. Since the child has red hair just like the mother, a recessive gene must also have passed from the father who has brown hair, because the brown colour gene is expressed over his red colour hair gene.

This allows each of the thousands of genes in an ovum to join the corresponding genes from a sperm when the chromosome pairs meet up at fertilization – you could call it love at first sight!

So you receive one set of chromosomes, the secret code of genes, from each parent. This means there are two genes (one from the father and one from the mother) for each characteristic that you inherit. One gene may be more influential than the other in the development of a specific trait of one of your parents. This more powerful gene is said to be dominant, while the less influential gene is recessive.

This variety of genes (dominant or recessive) and the characteristic it controls, such as brown or blue eye colour, is called an allele, and the terms gene and allele are used interchangeably. When two different alleles (dominant or recessive) are inherited, they are said to be heterozygous. When alleles are the same, they are termed homozygous. A dominant allele may be expressed when it is carried by only one of the chromosomes in a pair (the genotype is called a heterozygote) or both chromosomes (genotype is called homozygous dominant). A recessive allele, however, is only capable of expressing itself if a recessive allele is carried by both chromosomes in a pair (homozygous recessive).

Chromosomes

The chromosomal make-up of a cell is called its karyotype. Karyotyping is useful as it identifies if there is a chromosomal defect, which is too many or too few chromosomes, or whether chromosomes have lost or gained a part of another chromosome when the cell is dividing. It does not,

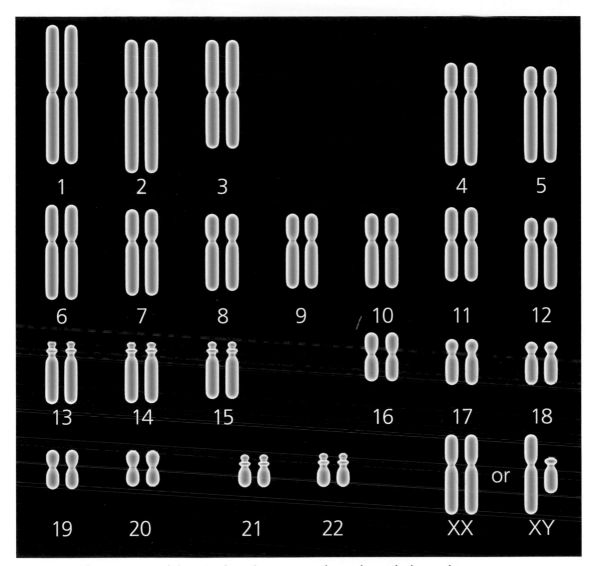

The 22 pairs of autosomes and the pair of sex chromosomes that make up the human karyotype.

however, provide information about the genetic make-up of a cell, called the genome, since this relates to the actual genes within the chromosomes.

Chromosomes vary in size and also according to the activities they control within the body. Two of the chromosomes, called sex chromosomes, are involved in the development of the sex organs of the embryo. Females have two X chromosomes and males have one X chromosome and one Y-chromosome. Their karyotypes are referred to as being XX and XY. The remaining 22 pairs are called autosomes, which are mostly concerned with the non-sexual activities of the body. The autosomes are numbered according to their size: pair 1 is the largest, pair 22 the smallest.

Each sperm produced by the male contains either an X or Y chromosome. When a sperm with an X chromosome fertilizes an ovum, the offspring is female and when a sperm with a Y chromosome fertilizes an ovum, the offspring is male. So it is the male who determines the sex of the baby.

What is a Mutation?

A mutation is a permanent change in the genetic material. When a gene mutates, it may produce a characteristic that is different from its original trait. Gene mutations in gametes may be passed on during reproduction. Some changes or mutations can cause serious or even lethal conditions.

There are three types of mutation: single gene disorders, chromosomal disorders and multifactorial disorders.

Single Gene Disorders

Single gene disorders are inherited in three clearly identifiable patterns, two of which are passed on by the autosomal chromosomes and are called autosomal dominant and autosomal recessive mutations. The third inheritance pattern is passed on by the sex chromosomes and this is called sex-linked inheritance. The majority of hereditary disorders are autosomal because there are 22 pairs of chromosomes and only one pair of sex chromosomes. Also, keeping the definitions of the terms dominant and recessive in mind, dominant genes produce abnormal traits in offsprings, even if only one parent has the gene. On the other hand, recessive genes do not produce abnormal traits unless both parents have the gene and pass it on to their offspring.

Predicting Inheritance

Dominant alleles are traditionally indicated by capitalized letters, and recessive alleles are abbreviated using lowercase letters. For a gene designated A, the possible genotypes are indicated by AA (homozygous dominant), Aa (heterozygous) or aa (homozygous recessive).

A simple box diagram, known as a punnett square, enables us to predict the probabilities that a given

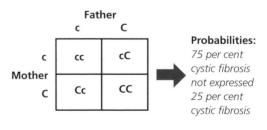

Carrier father

Carrier mother

c C

c C

c c
Affected son

c C
Carrier daughter

C c
Carrier son

C C
Unaffected daughter

The picture illustrates the potential offspring of two unaffected parents, each with an altered recessive gene (c) on an autosomal chromosome. Each offspring will have one in four chance of being affected (cc) with cystic fibrosis and a two in four chance of being a carrier (Cc or cC) of the condition.

	Father	
	c	C
c	cc	cC
C	Cc	CC

Mother (left label)

Probabilities:
75 per cent cystic fibrosis not expressed 25 per cent cystic fibrosis

Genetics and Inheritance

child will have particular characteristics (called phenotypes). In a punnett square (see diagram on the left), the two potential maternal alleles in ova are listed along the horizontal axes and the two potential paternal alleles in sperm along the vertical axes. The possible combination of alleles a child can inherit at fertilization are shown in the small boxes.

Autosomal Recessive Inheritance

Male and female offspring are affected equally as this is an inheritance of an disorder of the autosomal chromosomes.

- If both parents are unaffected, but heterozygous (carriers) for the trait (Cc), each of their offspring has a one in four chance of being affected (cc).
- If both parents are affected (cc), then all of the offspring will be affected (cc).
- If one parent is affected (cc) and the other is not a carrier (CC), all of their offspring will be unaffected, but will carry the altered gene (Cc).
- If one parent is affected (cc) and the other is a carrier (Cc), each of the offspring will have a one in two chance of being affected (cc) and therefore a one in two chance of being unaffected (Cc).

Autosomal Recessive Disorders

Cystic Fibrosis (CF)

This is the most common inherited condition, and is characterized by the secretion of thick mucus in the lungs and gastrointestinal tract that may obstruct airways and the pancreatic duct. The CF gene is found on chromosome 7.

Phenylketonuria (PKU)

This disorder is distinguished by a build-up of the essential amino acid phenylalanine in tissues. Young children in particular are at risk of the consequences of this problem (that is because phenylalanine accumulation interferes with brain uptake of other amino acids, and so slows brain development). The PKU gene is located on chromosome 12.

Familial Hypercholesterolemia (FH)

In this condition, there is an excessively high lipid concentration in the blood, which promotes the development of fatty (atheromatous) deposits in blood vessels. These lead to a reduction in blood flow to the cells that are supplied by these affected vessels. The gene for FH is found on chromosome 19.

Other examples of autosomal recessive inheritance include straight hair, blond hair, red hair, freckles, earlobes attached to the face, lack of A, B surface antigens on the red blood cell membrane (i.e., type O blood group), lack of rhesus antigens on the red blood cell membrane (i.e., Rhesus negative blood group), albinism and an inability to roll the tongue into a U-shape.

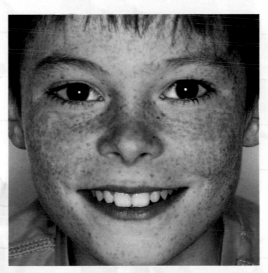

Freckles are a harmless example of autosomal recessive inheritance.

Autosomal Dominant Inheritance

- Male and female offspring are affected equally regarding the inherited pattern.
- If only one parent is affected and is a heterozygote (one unaffected blue box, one affected clear box, see diagram below), the children have a 50 per cent chance of being affected (i.e., they have one unaffected blue box and one clear affected box).
- If both parents are affected as homozygous dominants (i.e., both parents only have clear boxes) all the children will be affected (i.e., they have only clear boxes).
- If both parents are affected as heterozygous dominants (i.e., both parents have a clear box and

a blue box), then the chances are that three out of four children will be affected (i.e., two children have a clear affected box and a blue unaffected box, one child will have two clear affected boxes, and one child has a normal genotype of two recessive genes or two blue boxes).

In summary, two parents with a heterozygous dominant genotype will always have a one in four chance of producing a child with a recessive genotype. So if your parents both have brown eyes and you have blue, do not worry about your parentage, they are your parents – it just means that are both carriers of the blue recessive gene.

The picture illustrates the potential offspring of a parent with recessive normal genes (blue square box) and a parent with a mutated dominant gene (clear square box). Note with each pregnancy with these parental genotypes there is always a 50 per cent chance that the offspring will be affected (i.e., one clear and one blue box).

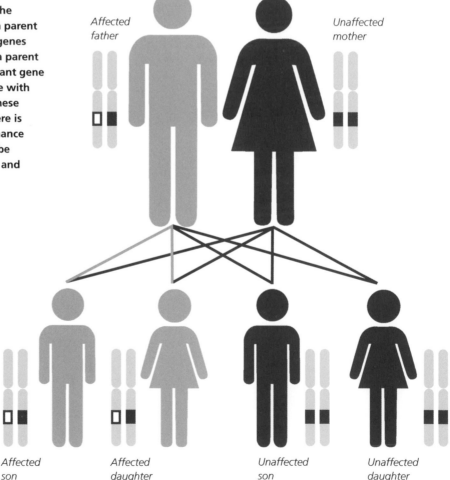

Affected father

Unaffected mother

Affected son

Affected daughter

Unaffected son

Unaffected daughter

Autosomal Dominant Conditions

Retinoblastoma is a rare eye tumour that originates in the retina of one or both eyes. In about 50 per cent of cases, the condition is acquired in the early years. The other 50 per cent are born with the tumour. The gene for the tumour is located on chromosome 13.

Huntington's disease (or Huntington's chorea) is a neurological disorder, but for some unknown reason its expression typically does not occur until between 40 to 50 years of age. So an individual may be totally unaware of its inheritance until long after they have had children of their own. This means they may have passed on this condition to their children. This disorder is characterized by a deficiency of the neurotransmitter gamma aminobutyric acid (GABA) from neurological areas involved in the control of movement. The gene is found on chromosome 4.

Other examples of autosomal dominant inheritance include curly hair, the absence of freckles, ear lobes that hang freely from the face, an ability to roll your tongue in a U-shape, the presence of the rhesus factor on the red blood cells (i.e., Rhesus positive blood group) and Marfan's syndrome.

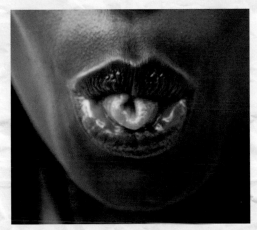

The ability to roll your tongue is an example of autosomal dominant inheritance.

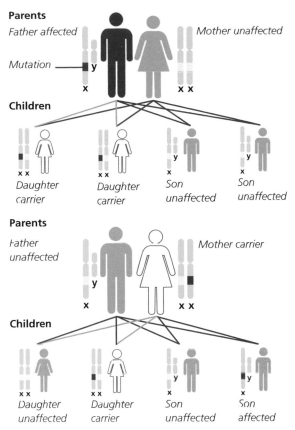

X-linked recessive inheritance

Sex-Linked Inheritance

Some genetic disorders arise through genes found on the sex chromosomes and are termed sex-linked. The Y-chromosome is not known to possess disease-causing genes, so the terms X-linked and sex- linked are interchangeable. Because females receive two X chromosomes (one from the father and one from the mother), they can be homozygous for a diseased allele, homozygous for a normal allele, or heterozygous. However, males have only one X chromosome, so even a single X-linked recessive gene can cause a disease. That is why they are either affected or not affected (see above). In comparison, the female needs two copies of the disease recessive gene (and only one copy of disease dominant gene). Therefore, since males only have one X chromosome they are more commonly affected by X- linked recessive

Sex-Linked Recessive Conditions

Duchenne's muscular dystrophy causes a loss of the muscle protein called dystrophin, which results in progressive wastage of muscle. Women who have the condition would have to be homozygous for the gene and so would have to have inherited an affected X chromosome from the father (who would have had this progressive condition himself, and so would be unlikely to have children). As a result, a woman might have one copy and so be a carrier of the muscular dystrophy allele, but is unlikely to have two copies and so have the condition. Nevertheless, a handful of cases of women with Duchenne's muscular dystrophy are known worldwide.

Haemophilia A results from a loss of the liver enzyme necessary for the synthesis of clotting factor VIII. The condition has been known for many years to primarily be a condition of boys, although a few incidences in girls have been documented. Haemophilia is now controllable (through administration of extrinsic Factor VIII) and so does not now place a reproductive restriction on the inherited genotype. Thus, it should be feasible for the man with haemophilia to father a daughter who, if she also inherited an affected X chromosome from her carrier mother, would be homozygous for the condition. The mutated allele has a low frequency within the population. This means that the partner of the father is unlikely to be carrying the gene and so the condition remains predominantly a problem in males.

Severe combined immunodeficiency syndrome (SCIDS) is a recessive condition in which there is a deficiency of proteins involved in the immune system. This results in a defective, and often deficient, population of T-lymphocytes, and sometimes of B-lymphocytes, leading to a severely compromised immune system. SCIDS is often referred to as "Boy in the Bubble disease" after a highly-publicized case in the 1970s when a child was physically isolated from his environment, but died as a teenager when that "protective bubble" was removed.

X-Linked Dominant Inheritance

- A person with an abnormal trait typically will have one affected parent.
- If a father has an X-linked dominant disorder, all his daughters and none of his sons will be affected.
- If the mother has an X-linked dominant disorder, there is a 50 per cent chance that each of her children will be affected.
- Evidence of an inherited trait most often appears in the family history.
- X-linked dominant disorders are commonly lethal in males.
- The family history may show miscarriages and the predominance of female offspring.

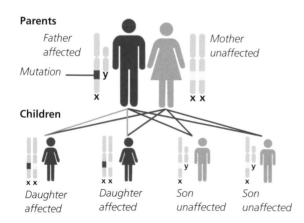

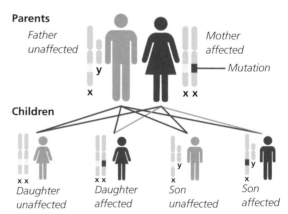

X-linked dominant inheritance

Chromosomal Disorders

Chromosomal disorders occur through deviations in either the inheritance of an abnormal number of chromosomes or through an abnormal structure of chromosome being passed on.

The inheritance of a numerical disturbance will occur if a chromosome pair fail to separate at meiosis when the gametes are formed. Failure of separation is referred to as a nondisjunction, and this can result in gametes with two of the same chromosome appearing in some gametes. If this gamete takes part in fertilization, then the resultant zygote will have three of the chromosome, referred to as trisomy – two copies will have been inherited from the gamete with nondisjunction, and one as normal from the other gamete. However, if nondisjunction leaves both copies of a chromosome in one gamete, then it follows that there will also be a gamete formed that lacks the chromosome. If that gamete takes part in fertilization then the resultant zygote will have just one copy of the chromosome and this is referred to as monosomy – the one copy will have been inherited from one partner as normal, the other gamete will not have a copy due to nondisjunction.

Trisomies and monosomies mean that an embryo has respectively either a great excess of alleles, or only has one set of alleles, associated with the chromosome involved. This imbalance appears to be detrimental to development if the chromosome is one of the autosomes, especially for monosomies, but babies with trisomies 13, 18 and 21 do occur, though trisomy 13 and 18 usually result in miscarriages and, if they survive birth, the child normally dies early on in infancy. However, viable trisomies and monosomies are observed for the sex chromosomes.

Nondisjunction can occur in the formation of either spermatozoa or oocytes, but most often arise during the formation of the oocyte, and their incidence seems to be related to advancing maternal age. For example, the incidence of trisomy 21 in Down syndrome is about 1 in 700 to 1,000 births overall, but if the mum is over 35 years old then the risk triples to 3 in 1,000.

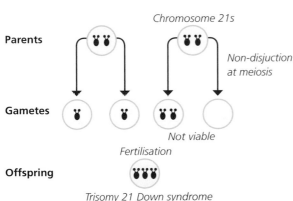

This diagrams show the nondisjunction of autosome chromosomes (left) and the nondisjunction of sex chromosomes.

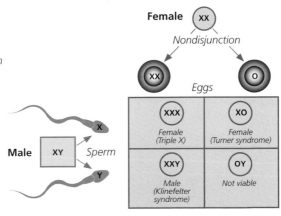

This diagram illustrates normal disjunction and nondisjunction of an ovum. The result is one trisomic cell and one monosomic (non-viable) cell.

Inherited Conditions

Trisomies

Trisomy 21 is usually known as Down syndrome (after the clinician who first described it) and it is the only viable autosomal trisomy. Trisomy 21 produces a number of developmental anomalies because now the child has three genes instead of two competing for their expression. These anomalies include various levels of learning disability, distinctive facial characteristics, especially of the forehead and eyes, an open mouth with a protruding tongue, cardiovascular anomalies, especially of heart structure, a third fontanelle of the skull and a single palmar crease (see below).

Klinefelter's

Klinefelter's syndrome is observed in males, in whom there is an extra X-chromosome, so these individuals are XXY. The result is that the testes are small and androgen secretion is low. Sperm may be absent, though some people with the syndrome have fathered children.

Trisomy

Trisomy of the X-chromosome is referred to as Triple X syndrome and it is frequently found within a population, but individuals usually do not have distinguishing characteristics. In the absence of a Y chromosome, triple-X people are always female. Although some of these women have menstrual difficulties, many are fertile.

Monosomies

Cells normally require at least two copies of each chromosome and so monosomic foetuses do not go full-term. The exception is Turner's syndrome in which individuals are XO, so only have one X chromosome. Even then, only two out of every hundred survive full-term and are born female, as no Y-chromosome is present. Development of the gonads, and body growth, are slowed, and there may be a range of systemic problems.

Should a zygote inherit just a Y chromosome (i.e., YO) then it does not develop. Apparently, at least one copy of an X chromosome is a minimum requirement for foetal development.

Fragmentation of Chromosomes

Translocation is the movement of a part of a chromosome. It occurs when chromosomes break and rejoin in an abnormal arrangement. When the rearrangement preserves the normal amount of genetic material, there are no visible abnormalities. This is called balanced translocation. However, when the rearrangement changes the genetic expression, visible abnormalities usually occur.

The children of parents with these unbalanced translocations may have serious, chromosomal anomalies, such as partial monosomies or partial trisomies. Parental age doesn't seem to be a factor for the process of translocation.

Transplanting Genes

Gene transplantation (called transgenics) is an inspiring development which could help the prevention or correction of genetic disorders. It involves incorporating the DNA of a "healthy" allele into a carrier organism (called a vector) such as a virus. These carriers are used because viruses specifically target specific host tissues, which might be in need of the transplanted genes. The viral DNA (with its donor "healthy DNA") becomes integrated into the DNA of the recipient "diseased" cell. If the transplanted donor allele is a dominant form, then this will correct the genetic recessive disorder. To be effective in correcting an established disorder, a normal allele must be transplanted into the majority of the affected cells in already differentiated recipients' tissues. One way is to transplant the healthy gene into stem cells (i.e., cells from which new tissue cells are produced), then introduce the stem cells (and the

healthy DNA) into a person with a genetic disorder. In humans, this technique has been used to treat leukaemia in young babies and, currently, attempts are being made to introduce sufficient genes to produce significant health improvements in people with conditions such as cystic fibrosis and diabetes mellitus.

Genetic Fingerprinting

The ability of scientists to replicate a DNA sample many times over, and determine the genetic coding of the sample, has enabled the analysis of minute traces of DNA. This has helped crime investigations (hence the term "DNA fingerprinting"), the identification of related species of plant or animal (both living and extinct!) and in the debate about the evolutionary links of human cultures.

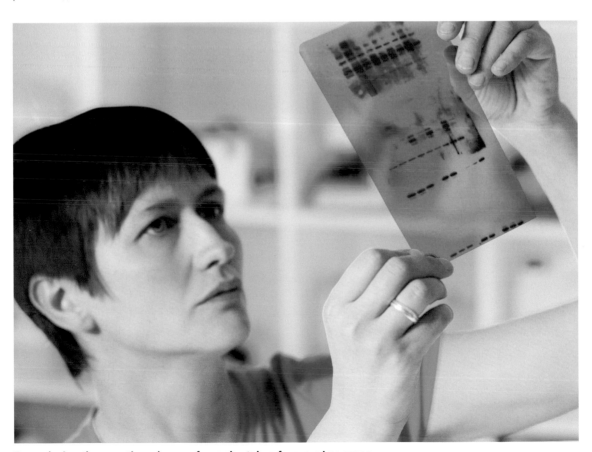

By analyzing the genetic make-up of samples taken from a crime scene, detectives can match them to the DNA from suspects.

Index

Picture Credits

Dreamstime.com: Addictivex 6; Radub85 8; Alxhar 11; Sebastian Kaulitzki 14l, 14r, 63; Kittipong Jirasukhanont 15 tl, 58; Science Pics 15tr, 45, 46b, 106, 160; Designua 16, 17, 68t, 81, 114, 116, 158t; Kateryna Kon 19, 95, 137; Nicolas Fernandez 20t; Puntasit Choksawatkidorn 20b; Andegraund548 22; Christian Weiß 23; Kazakov Alexey 26, Khotcharak Siriwong 28; Tyler Olsen 30l; Keantian 30r; Pimonpim Tangosol 31; Patricia Hofmeester 32t; Anna Bocharova 33; Cornelius20 36; Newartgraphics 38; Tigatelu 39t; Alicia Gonzalez 39b; Stihii 41; Dannyphoto80 44, 48r, 55; Jubalharshaw19 46t, 47; Starast 48; Bluringmedia 49; Suriya Siritam 50b; Peter Junaidy 51, 155; Anton Starikov 53; la64 56; Andreus 59, 60, 141; Molekuul 61; Luk Cox 64; Alila07 10l, 65, 68b, 77, 90, 91t, 98, 115, 129, 130; Tacio Philip Sansonovski 66; Eveleen007 67; Jurica Vukovic 69; Wavebreakmedia Ltd 71l, 112; Todsaporn Bunmuen 74t, Snapgalleria 76; Maxim Sivyi 79; Georgerudy 81r; Markuk97 82t; Jonmilnes 82b; Doethio 83; Yobro10 84; Alessandro Melis 85t; Elenabsl 86, 144; Atman 88b; Rob3000 87, 102, 174; Alexander Ishchenko 88t; Lightmemorystock 91b, Guniita 10r, 93, 118, 119, 121,146, 147; Josha42 94; Legger 96, 100, 104b, 117, 131; Maglyvi 97; Vesna Njagulj 101; Brett Critchley 104t; Lukaves 107; Piotr Marchinski 108; Miraw5 109t; Tomnex 109b; Zbynek Pospisil 110b; Zlikovec 120; Bluezace 123; Dny3dcom 128t; Czuber 128b; Turhanerbas 132; Marochkina 135, 138; Martin Hatch 136; Andegraund548 139; Jaeeho 140; Jurica Vukovic 142t; Stratum 148; Airborne77 149; Skypixel 150; Marcogovel 152; Merzzie 154; Alexluengo 156; Zuzanaa 157; Pattarawit Chompipat 159; Magicmine 161; Shubhangi Kene 162; Anolkil 164; Anatoly Shevkunov 165; Kazakov Alexey 166; Irina Rumiantseva 167; Kristopher Strach 170; Iryna Timonina 172; Pglazar 175l; Luk Cox 177; 178; Gunold 179; 180; Joshua Wanyama 181l; 181r; 182; Olesia Bilkei 184; Luchschen 185.

Shutterstock: MriMan 27; eranicle 35; Jose Luis Calvo 42.

Wikimedia Commons: 15b; Afunguy 18b; Rollroboter 21r; Michael Goodyear 24t; National Cancer Institute 24b; BruceBlaus 34; Plainpaper 70; CDC/Dr Terrence Tumpey/Cynthia Goldsmith 72; National Institutes of Health (NIH) 73; Manu5 99t; 124; James Heilman, MD 163l; OpenStax College 169; Jonnymccullagh 175r.

Illustration Credit

Daniel Limon (Beehive Illustration)

Every effort has been made to contact copyright holders. The publishers will be pleased to make good any omissions or rectify any mistakes brought to their attention at the earliest opportunity.

Publishing Director *Trevor Davies*
Production Controller *Katie Jarvis*

For Tall Tree Ltd
Editors *Emma Marriott and Jon Richards*
Assistant editor *Lauren Clancy*
Designers *Gary Hyde, Ben Ruocco and Jonathan Vipond*